Friedrich Max Müller

Lectures on the Science of Religion

with a paper on Buddhist nihilism, and a translation of the Dhammapada

Friedrich Max Müller

Lectures on the Science of Religion
with a paper on Buddhist nihilism, and a translation of the Dhammapada

ISBN/EAN: 9783337262860

Printed in Europe, USA, Canada, Australia, Japan

Cover: Foto ©Lupo / pixelio.de

More available books at **www.hansebooks.com**

LECTURES

ON THE

SCIENCE OF RELIGION;

WITH A PAPER ON

BUDDHIST NIHILISM,

AND A TRANSLATION OF THE

DHAMMAPADA OR "PATH OF VIRTUE."

BY

MAX MÜLLER, M. A.

FELLOW OF ALL-SAINTS' COLLEGE, OXFORD, CORRESPONDANT DE L'INSTITUT DE FRANCE, AUTHOR OF "LECTURES ON THE SCIENCE OF LANGUAGE," "CHIPS FROM A GERMAN WORKSHOP," ETC.

NEW YORK:
SCRIBNER, ARMSTRONG, AND CO.,
SUCCESSORS TO
CHARLES SCRIBNER AND CO.
1872.

RIVERSIDE, CAMBRIDGE:
STEREOTYPED AND PRINTED BY
H. O. HOUGHTON AND COMPANY

CONTENTS.

	PAGE
LECTURES ON THE SCIENCE OF RELIGION.	
First Lecture	3
Second Lecture	29
Third Lecture	54
Fourth Lecture	100
BUDDHIST NIHILISM	131
BUDDHA'S DHAMMAPADA, OR "PATH OF VIRTUE."	
Introduction	151

CHAPTER I.
The Twin-verses 193

CHAPTER II.
On Reflection 200

CHAPTER III.
Thought 203

CHAPTER IV.
Flowers 207

CHAPTER V.
The Fool 211

CHAPTER VI.
The Wise Man 215

CHAPTER VII.
The Venerable 219

CHAPTER VIII.
The Thousands 223

CHAPTER IX.
Evil 227

CHAPTER X.
Punishment 230

CONTENTS.

CHAPTER XI.
Old Age 235

CHAPTER XII.
Self 238

CHAPTER XIII.
The World 241

CHAPTER XIV.
The Awakened (Buddha) 244

CHAPTER XV.
Happiness 249

CHAPTER XVI.
Pleasure 253

CHAPTER XVII.
Anger 256

CHAPTER XVIII.
Impurity 259

CHAPTER XIX.
The Just 264

CHAPTER XX.
The Way 268

CHAPTER XXI.
Miscellaneous 272

CHAPTER XXII.
The Downward Course 276

CHAPTER XXIII.
The Elephant 279

CHAPTER XXIV.
Thirst 282

CHAPTER XXV.
The Bhikshu (Mendicant) 288

CHAPTER XXVI.
The Brâhmana 293

LECTURES ON
THE SCIENCE OF RELIGION.

By MAX MÜLLER,

PROFESSOR OF COMPARATIVE PHILOLOGY IN THE UNIVERSITY OF OXFORD, ETC.

FIRST LECTURE.

DELIVERED AT THE ROYAL INSTITUTION, FEB. 19, 1870.

WHEN I undertook for the first time to deliver a course of lectures in this Institution, I chose for my subject the *Science of Language*. What I then had at heart was to show to you, and to the world at large, that the comparative study of the principal languages of mankind was based on principles sound and scientific, and that it had brought to light results which deserved a larger share of public interest than they had as yet received. I tried to convince, not only scholars by profession, but historians, theologians, and philosophers, nay everybody who had once felt the charm of gazing inwardly upon the secret workings of his own mind, veiled and revealed as they are in the flowing forms of language, that the discoveries made by comparative philologists could no longer be ignored with impunity; and I submitted that after the progress achieved in a scientific study of the principal branches of the vast realm of human speech, our new science, the Science of Language, might claim by right its seat at the round-table of the intellectual chivalry of our age.

Such was the goodness of the cause I had then to defend, that, however imperfect my own pleading, the verdict of the public has been immediate and almost unanimous. During the years that have elapsed since the delivery of my first course of lectures, the Science of Language has had its full share of public recognition. Whether we look at the number of books that have been published for the advancement and elucidation of our science, or at the excellent articles in the daily, weekly, fortnightly, monthly, or quarterly reviews, or at the frequent notices of its results scattered about in works on philosophy, theology, and ancient history, we may well rest satisfied. The example set by France and Germany, in founding chairs of Sanskrit and Comparative Philology, has been followed of late in nearly all the universities of England, Ireland, and Scotland. We need not fear for the future of the Science of Language. A career so auspiciously begun, in spite of strong prejudices that had to be encountered, will lead on from year to year to greater triumphs. Our best public schools, if they have not done so already, will soon have to follow the example set by the universities. It is but fair that school-boys who are made to devote so many hours every day to the laborious acquisition of languages, should now and then be taken by a safe guide to enjoy from a higher point of view that living panorama of human speech which has been surveyed and carefully mapped out by patient explorers and bold discoverers: nor is there any longer an excuse why, even in the most elementary lessons, nay I should say, why more particularly in these elementary lessons, the dark and dreary passages of Greek and Latin, of French and German grammar, should not be lighted up by the electric light of Comparative Philology. When last year I travelled in Germany I found that lectures on Comparative Philology are now attended in the universities by all who study Greek and Latin. At Leipzig alone the lectures of the professor of Sanskrit were attended by more than fifty under-graduates, who first acquire that amount of knowledge of Sanskrit which is absolutely necessary before entering upon a study of Comparative Grammar. The introduction of Greek into the universities of Europe in the fifteenth century could hardly have caused a greater revolution than the discovery of Sanskrit and the study of Comparative Philology in the nineteenth century. Very few indeed now take their degree of Master of Arts in Germany, or would be allowed to teach at a public school, without having been examined in the principles of Comparative Philology, nay in the elements of Sanskrit Grammar. Why should it be different in England? The intellectual fibre, I know, is not different in the youth of England and in the youth of Germany, and if there is but a fair field and no favor, Comparative Philology, I feel convinced, will soon hold in England, too, that place which it ought to hold at every public school, in every university, and in every classical examination.

In beginning to-day a course of lectures on the *Science of Religion*,—or I should rather say on some preliminary points that have to be settled before we can enter upon a truly scientific study of the religions of the world,—I feel as I felt when first pleading in this very place for the Science of Language.

I know that I shall have to meet determined antagonists who will deny the possibility of a scientific treatment of religions as they denied the possibility of a scientific treatment of languages. I foresee even a far more serious conflict with familiar prejudices and deep-rooted convictions; but I feel at the same time that I am prepared to meet my antagonists; and I have such faith in their honesty of purpose, that I doubt not of a patient and impartial hearing on their part, and of a verdict influenced by nothing but by the evidence that I shall have to place before them.

In these our days it is almost impossible to speak of religion without giving offense either on the right or on the left. With some, religion seems too sacred a subject for scientific treatment: with others it stands on a level with alchemy and astrology, a mere tissue of errors or hallucinations, far beneath the notice of the man of science. In a certain sense, I accept both these views. Religion is a sacred subject, and whether in its most perfect or in its most imperfect form, it has a right to our highest reverence. No one — this I can promise — who attends these lectures, be he Christian or Jew, Hindu or Mohammedan, shall hear his own way of serving God spoken of irreverently. But true reverence does not consist in declaring a subject, because it is dear to us, to be unfit for free and honest inquiry; far from it! True reverence is shown in treating every subject, however sacred, however dear to us, with perfect confidence; without fear and without favor; with tenderness and love, by all means, but, before all, with an unflinching and uncompromising loyalty to truth. I also admit that religion has stood in former ages, and stands even in our own age, if we look abroad, aye, even if we look into some dark places at home, on a level with alchemy and astrology; but for the discovery of truth there is nothing so useful as the study of errors, and we know that in alchemy there lay the seed of chemistry, and that astrology was more or less a yearning and groping after the true science of astronomy.

But although I shall be most careful to avoid giving

offense, I know perfectly well that many a statement I shall have to make, and many an opinion I shall have to express, will sound strange and startling to some of my hearers. The very title of the Science of Religion jars on the ears of many persons, and a comparison of all the religions of the world, in which none can claim a privileged position, must seem to many reprehensible in itself, because ignoring that peculiar reverence which everybody, down to the mere fetich worshipper, feels for his *own* religion and for his *own* God. Let me say then at once that I myself have shared these misgivings, but that I have tried to overcome them, because I would not and could not allow myself to surrender either what I hold to be the truth, or what I hold still dearer than the truth, the right tests of truth. Nor do I regret it. I do not say that the Science of Religion is all gain. No; it entails losses, and losses of many things which we hold dear. But this I will say, that, as far as my humble judgment goes, it does not entail the loss of anything that is essential to true religion, and that if we strike the balance honestly, the gain is immeasurably greater than the loss.

One of the first questions that was asked by classical scholars when invited to consider the value of the Science of Language, was, "What shall we gain by a comparative study of languages?" Languages, it was said, are wanted for practical purposes, for speaking and reading; and by studying too many languages at once, we run the risk of losing the firm grasp which we ought to have on the few that are really important. Our knowledge, by becoming wider, must needs, it was thought, become shallower, and the gain, if there is any, in knowing the structure of dialects which have never produced any literature at all, would certainly be outweighed by the loss in accurate and practical scholarship.

If this could be said of a comparative study of languages, with how much greater force will it be urged against a comparative study of religions! Though I do not expect that those who study the religious books of Brahmans and Buddhists, of Confucius and Lao-tse, of Mohammed and Nanak, will be accused of cherishing in their secret heart the doctrines of those ancient masters, or of having lost the firm hold on their own religious convictions, yet I doubt whether the practical utility of wider studies in the vast field of the religions of the world will be admitted with greater readiness by professed theologians than the value of a knowledge of Sanskrit, Zend, Gothic, or Celtic for a thorough mastery of Greek and Latin, and for a real appreciation of the nature, the purpose, the laws, the growth and decay of language was admitted, or is even now admitted, by some of our most eminent professors and teachers.

People ask, What is gained by comparison? Why, all higher knowledge is gained by comparison, and rests on comparison. If it is said that the character of scientific research in our age is preëminently comparative; this really means that our researches are now based on the widest evidence that can be obtained, on the broadest inductions that can be grasped by the human mind. What can be gained by comparison? Why, look at the study of languages. If you go back but a hundred years and examine the folios of the most learned writers upon questions connected with language, and then open a book written by the merest tyro in Comparative Philology, you will see what can be gained, what has been gained, by the comparative method.

A few hundred years ago, the idea that Hebrew was the original language of mankind was accepted as a matter of course, even as a matter of faith, the only problem being to find out by what process Greek, or Latin, or any other language could have been developed out of Hebrew. The idea, too, that language was revealed, in the scholastic sense of that word, was generally accepted, although, as early as the fourth century, St. Gregory, the learned Bishop of Nyssa, had strongly protested against it. The grammatical frame-work of a language was either considered as the result of a conventional agreement, or the terminations of nouns and verbs were supposed to have sprouted forth like buds from the roots and stems of language; and the vaguest similarity in the sound and meaning of words was taken to be a sufficient criterion for testing their origin and their relationship. Of all this philological somnambulism we hardly find a trace in works published since the days of Humboldt, Bopp, and Grimm. Has there been any loss here? Has it not been pure gain? Does language excite admiration less because we know that, though the faculty of speaking is the work of Him who has so framed our nature, the invention of words for naming each object was left to man, and was achieved through the working of the human mind? Is Hebrew less carefully studied because it is no longer believed to be a revealed language sent down from heaven, but a language closely allied to Arabic, Syriac, and ancient Babylonian, and receiving light from these cognate, and in some respects more primitive languages, for the explanation of many of its grammatical forms, and for the exact interpretation of many of its obscure and difficult words? Is the grammatical articulation of Greek and Latin less instructive because, instead of seeing in the termination of nouns and verbs merely arbitrary signs to distinguish the singular from the plural, or the present from the future, we can now perceive an intelligible principle in the gradual production of formal out of the material elements of language? And are our etymologies less important because, instead of being suggested by superficial similarities, they are now based on honest historical and physiological research? Lastly, has our own language ceased to hold its own peculiar place? Is our love for our own native tongue at all impaired? Do men speak less boldly or pray less fervently in their own mother-tongue, because they know its true origin and its unadorned history; or because they have discovered that in all languages, even in the jargons of the lowest savages, there is order and wisdom; there is in them something that makes the world akin?

Why, then, should we hesitate to apply the comparative method, which has produced such great results in other spheres of knowledge, to a study of religion? That it will change many of the views commonly held about the origin, the character, the growth, and decay of the religions of the world, I do not deny; but unless we hold that fearless progression in new inquiries, which is our bounden duty and our honest pride in all other branches of knowledge, is dangerous in the study of religions, unless we allow ourselves to be frightened by the once famous dictum, that whatever is new in theology is false, this ought to be the very reason why a comparative study of religions should no longer be neglected or delayed.

When the students of Comparative Philology boldly adopted Goethe's paradox, "*He who knows one language, knows none;*" people were startled at first, but they soon began to feel the truth which was hidden beneath the paradox. Could Goethe have meant that Homer did not know Greek, or that Shakespeare did not know English, because neither of them knew more than his own mother-tongue? No! what was meant was that neither Homer nor Shakespeare knew what that language really was which he handled with so much power and cunning. Unfortunately the old verb " to can," from which " canny " and " cunning," is lost in English, otherwise we should be able in two words to express our meaning, and to keep apart the two kinds of knowledge of which we are here speaking. As we say in German *können* is not *kennen*, we might say in English *to can*, that is to be cunning, is not *to ken*, that is to know; and it would then become clear at once, that the most eloquent speaker and the most gifted poet, with all their command of words and skillful mastery of expression, would have but little to say if asked what language really is! The same applies to religion. *He who knows one, knows none.* There are thousands of people whose faith is such that it could move mountains, and who yet, if they were asked what religion really is, would remain silent, or would speak of outward tokens rather than of the inward nature, or of the faculty of faith.

It will easily be perceived that religion means at least two very different things. When we speak of the Jewish, or the Christian, or the Hindu religion, we mean a body of doctrines handed down by tradition, or in canonical books, and containing all that constitutes the faith of Jew, Christian, or Hindu. Using religion in that sense, we may say that a man has changed his religion, that is, that he has adopted the Christian instead of the Brahmanical body of religious doctrines, just as a man may learn to speak English instead of Hindustani. But religion is also used in a different sense. As there is a faculty of speech, independent of all the historical forms of language, so we may speak of a faculty of faith in man, independent of all historical religions. If we say that it is religion which distinguishes man from the animal, we do not mean the Christian or Jewish religions only; we do not mean any special religion, but we mean a mental faculty, that faculty which, independent of, nay in spite of sense and reason, enables man to apprehend the Infinite under varying disguises. Without that faculty, no religion, not even the lowest worship of idols and fetiches, would be possible; and if we will but listen attentively, we can hear in all religions a groaning of the spirit, a struggle to conceive the inconceivable, to utter the unutterable, a longing after the Infinite, a love of God. Whether the etymology which the ancients gave of the Greek word ἄνθρωπος, man, be true or not (they derived it from ὁ ἄνω ἀθρῶν, he who looks upward): certain it is that what makes man to be man, is that he alone can turn his face to heaven; certain it is that he alone yearns for something that neither sense nor reason can supply.

If then there is a philosophical discipline which examines into the conditions of sensuous perception, and if there is another philosophical discipline which examines into the conditions of rational conception, there is clearly a place for a third philosophical discipline that has to examine into the conditions of that third faculty of man, coördinate with sense and reason, the faculty of perceiving the Infinite, which is at the root of all religions. In German we can distinguish that third faculty by the name of *Vernuft*, as opposed to

Verstand, reason, and *Sinne*, sense. In English I know no better name for it than the faculty of faith, though it will have to be guarded by careful definition, and to be restricted to those objects only, which cannot be supplied either by the evidence of the senses, or by the evidence of reason. No simply historical fact can ever fall under the cognizance of faith.

If we look at the history of modern thought, we find that the dominant school of philosophy, previous to Kant, had reduced all intellectual activity to *one* faculty, that of the senses. "Nihil in intellectu quod non ante fuerit in sensu," "Nothing exists in the intellect but what has before existed in the senses," was their watch-word; and Leibnitz answered it epigrammatically, but most profoundly, "Nihil — nisi intellectus." "Yes, nothing but the intellect." Then followed Kant, who, in his great work written ninety years ago, but not yet antiquated, proved that our knowledge requires the admission of two independent faculties, the intuitions of the senses, and the categories, or, as we might call them, the necessities of reason. But satisfied with having established the independent faculty of reason, as coördinate with the faculty of sense, or, to use his own technical language, satisfied with having proved the possibility of apodictic judgments *à priori*, Kant declined to go further, and denied to the intellect the power of transcending the finite, the faculty of approaching the Divine. He closed the ancient gates through which man had gazed into Infinity, but, in spite of himself, he was driven, in his "Critique of Practical Reason," to open a side-door through which to admit the sense of the Divine. This is the vulnerable point in Kant's philosophy, and if philosophy has to explain what is, not what ought to be, there will be and can be no rest till we admit, what cannot be denied, that there is in man a third faculty, which I call simply the faculty of apprehending the Infinite, not only in religion, but in all things; a power independent of sense and reason, a power in a certain sense contradicted by sense and reason, but yet, I suppose, a very real power, if we see how it has held its own from the beginning of the world, how neither sense nor reason have been able to overcome it, while it alone is able to overcome both reason and sense.

According to the two meanings of the word religion, then, the science of religion is divided into two parts; the former, which has to deal with the historical forms of religion, is called *Comparative theology;* the latter, which has to explain the conditions under which religion, in its highest or lowest form, is possible, is called *Theoretic theology.*

We shall at present have to deal with the former only; nay, it will be my object to show that the problems which chiefly occupy theoretic theology, ought not to be taken up till all the evidence that can possibly be gained from a comparative study of the religions of the world has been fully collected, classified, and analyzed.

It may seem strange that while theoretical theology, or the analysis of the inward and outward conditions under which faith is possible, has occupied so many thinkers, the study of comparative theology has never as yet been seriously taken in hand. But the explanation is very simple. The materials on which alone a comparative study of the religions of mankind could have been founded were not accessible in former days, while in our own days they have come to light in such profusion as almost to challenge these more comprehensive inquiries in a voice that cannot be disobeyed.

It is well known that the Emperor Akbar had a passion for the study of religions, so that he invited to his court Jews, Christians, Mohammedans, Brahmans, and Fire-worshippers, and had as many of their sacred books as he could get access to, translated for his own study. Yet, how small was the collection of sacred books that even an emperor of India could command not more than two hundred and fifty years ago, compared to what may now be found in the library of every poor scholar! We have the original text of the Veda, which neither the bribes nor the threats of Akbar could extort from the Brahmans. The translation of the Veda which he is said to have obtained, was a translation of the so-called Atharva-veda, and comprised most likely the Upanishads only, mystic and philosophical treatises, very interesting, very important in themselves, but as far removed from the ancient poetry of the Veda as the Talmud is from the Old Testament, as Sufiism is from the Koran. We have the Zendavesta, the sacred writings of the so-called fire-worshippers, and we possess the translation of it, far more complete and far more correct than any that the Emperor Akbar could have obtained. The religion of Buddha, certainly in many respects more important than either Brahmanism, or Zoroastrianism, or Mohammedanism, is never mentioned in the religious discussions that took place one evening in every week at the imperial court of Delhi. Abufazl, it is said, the minister of Akbar, could find no one to assist him in his inquiries respecting Buddhism. We possess the whole sacred canon of the Buddhists in various languages, in Pali, in Sanskrit, in Burmese, Siamese, Tibetan, Mongolian, and Chinese, and it is our fault entirely, if as yet there is no complete translation in any European tongue of this important collection of sacred books. The ancient religions of China again, that of Confucius and that of Laotse, may now be studied in excellent translations of their sacred books by anybody interested in the ancient faith of mankind.

But this is not all. We owe to missionaries particularly, careful accounts of the religious belief and worship among tribes far lower in the scale of civilization than the poets of the Vedic hymns, or the followers of Confucius. Though the belief of African and Melanesian savages is more recent in point of time, it represents an earlier and far more primitive phase in point of growth, and is therefore as instructive to the student of religion as the study of uncultivated dialects has proved to the student of language.

Lastly, and this, I believe, is the most important advantage which we enjoy as students of the history of religion, we have been taught the rules of critical scholarship. No one would venture nowadays, to quote from any book, whether sacred or profane, without having asked these simple and yet momentous questions: When was it written? Where? and by whom? Was the author an eye-witness, or does he only relate what he has heard from others? And if the latter, were his authorities at least contemporaneous with the events which they relate, and were they under the sway of party feeling or any other disturbing influence? Was the whole book written at once, or does it contain portions of an earlier date; and if so, is it possible for us to separate these earlier documents from the body of the book?

A study of the original documents on which the

principal religions of the world profess to be founded, carried out in this spirit, has enabled some of our best living scholars to distinguish in each religion between what is really ancient and what is comparatively modern; what was the doctrine of the founders and their immediate disciples, and what were the afterthoughts and, generally, the corruptions of later ages. A study of these later developments, of these later corruptions, or, it may be, improvements, is not without its own peculiar charms, and full of practical lessons; yet, as it is essential that we should know the most ancient forms of every language, before we proceed to any comparisons, it is indispensable that we should have a clear conception of the most primitive form of every religion before we proceed to determine its own value, and to compare it with other forms of religious faith. Many an orthodox Mohammedan, for instance, will relate miracles wrought by Mohammed; but in the Koran Mohammed says distinctly, that he is a man like other men. He disdains to work miracles, and appeals to the great works of Allah, the rising and setting of the sun, the rain that fructifies the earth, the plants that grow, and the living souls that are born into the world, — who can tell whence? — as the real signs and wonders in the eyes of a true believer.

The Buddhist legends teem with miserable miracles attributed to Buddha and his disciples — miracles which in wonderfulness certainly surpass the miracles of any other religion: yet in their own sacred canon a saying of Buddha's is recorded, prohibiting his disciples from working miracles, though challenged by the multitudes who required a sign that they might believe. And what is the miracle that Buddha commands his disciples to perform? "Hide your good deeds," he

2

says, " and confess before the world the sins you have committed."

Modern Hinduism rests on the system of caste as on a rock which no arguments can shake; but in the Veda, the highest authority of the religious belief of the Hindus, no mention occurs of the complicated system of castes, such as we find it in Manu: nay, in one place, where the ordinary classes of the Indian, or any other society, are alluded to, namely, the priests, the warriors, the citizens, and the slaves, all are represented as sprung alike from Brahman, the source of all being.

It would be too much to say that the critical sifting of the authorities for a study of each religion has been already fully carried out. There is work enough still to be done. But a beginning, and a very successful beginning, has been made, and the results thus brought to light will serve as a wholesome caution to everybody who is engaged in religious researches. Thus, if we study the primitive religion of the Veda, we have to distinguish most carefully, not only between the hymns of the Rig-veda on one side, and the hymns collected in the Sama-veda, Ya*g*ur-veda, and Atharva-veda on the other, but critical scholars would distinguish with equal care between the more ancient and the more modern hymns of the Rig-veda, as far as even the faintest indications of language, of grammar, or metre enable them to do so.

In order to gain a clear insight into the motives and impulses of the founder of the worship of Ahuramazda, we must chiefly, if not entirely, depend on those portions of the Zendavesta which are written in the Gatha dialect, a more primitive dialect than that of the rest of the sacred code of the Zoroastrians.

In order to do justice to Buddha, we must not mix

the practical portions of the Tripitaka, the Dharma, with the metaphysical portions, the Abhidharma. Both, it is true, belong to the sacred canon of the Buddhists; but their original sources lie in very different latitudes of religious thought.

We have in the history of Buddhism an excellent opportunity for watching the process by which a canon of sacred books is called into existence. We see here, as elsewhere, that during the life-time of the teacher, no record of events, no sacred code containing the sayings of the master was wanted. His presence was enough, and thoughts of the future, and more particularly of future greatness, seldom entered the minds of those who followed him. It was only after Buddha had left the world to enter into Nirvana, that his disciples attempted to recall the sayings and doings of their departed friend and master. At that time everything that seemed to redound to the glory of Buddha, however extraordinary and incredible, was eagerly welcomed, while witnesses who would have ventured to criticise or reject unsupported statements, or to detract in any way from the holy character of Buddha, had no chance of even being listened to. And when, in spite of all this, differences of opinion arose, they were not brought to the test by a careful weighing of evidence, but the names of "unbeliever" and "heretic" (nastika, pashanda) were quickly invented in India as elsewhere, and bandied backwards and forwards between contending parties, till at last, when the doctors disagreed, the help of the secular power had to be invoked, and kings and emperors convoked councils for the suppression of schism, for the settlement of an orthodox creed, and for the completion of a sacred canon. We know of King Asoka, the contemporary of Seleucus, sending his royal missive to the assembled elders, and telling them what to do, and what to avoid, warning them also in his own name of the apocryphal or heretical character of certain books which, as he thinks, ought not to be admitted into the sacred canon.

We here learn a lesson, which is confirmed by the study of other religions, that canonical books, though they furnish in most cases the most ancient and most authentic information within the reach of the student of religion, are not to be trusted implicitly, nay, that they must be submitted to a more searching criticism and to more stringent tests than any other historical books. For that purpose the Science of Language has proved in many cases a most valuable auxiliary. It is not easy to imitate ancient language so as to deceive the practiced eye of the grammarian, even if it were possible to imitate ancient thought that should not betray to the historian its modern origin. A forged book, like the Ezour Veda, which deceived even Voltaire, and was published by him as "the most precious gift for which the West was indebted to the East,' could hardly impose again on any Sanskrit scholar of the present day. This most precious gift from the East to the West, is about the silliest book that can be read by the student of religion, and all one can say in its defense is that the original writer never meant it as a forgery, never intended it for the purpose for which it was used by Voltaire. I may add that a book which has lately attracted considerable attention, "La Bible dans l'Inde," by M. Jacolliot, belongs to the same class of books. Though the passages from the sacred books of the Brahmans are not given in the original, but only in a very poetical French translation, no Sanskrit scholar would hesitate for one moment to say that they are forgeries, and that M. Jacolliot, the President of the Court of Justice at Chandernagore, has been deceived by his native teacher. We find many childish and foolish things in the Veda, but when we read the following line, as an extract from the Veda: —

> La femme c'est l'ame de l'humanite, —

it is not difficult to see that this is the folly of the nineteenth century, and not of the childhood of the human race. M. Jacolliot's conclusions and theories are such as might be expected from his materials.

With all the genuine documents for studying the history of the religions of mankind that have lately been brought to light, and with the great facilities which a more extensive study of Oriental languages has afforded to scholars at large for investigating the deepest springs of religious thought all over the world, a comparative study of religions has become a necessity. A science of religion, based on a comparison of all, or, at all events, of the most important religions of mankind, is now only a question of time. It is demanded by those whose voice cannot be disregarded. Its title, though implying as yet a promise rather than a fulfillment, has become more or less familiar in Germany, France, and America; its great problems have attracted the eyes of many inquirers, and its results have been anticipated either with fear or delight. It becomes the duty of those who have devoted their life to the study of the principal religions of the world in their original documents, and who value religion and reverence it in whatever form it may present itself, to take possession of this new territory in the name of true science, and thus to protect its sacred precincts from the inroads of mere babblers. Those who would use a comparative study of religions as a means for debasing Christianity by exalting the other religions of mankind, are to my mind as dangerous allies as those who think it necessary to debase all other religions in order to exalt Christianity. Science wants no partisans. I make no secret that true Christianity seems to me to become more and more exalted the more we appreciate the treasures of truth hidden in the despised religions of the world. But no one can honestly arrive at that conviction, unless he uses honestly the same measure for all religions. It would be fatal for any religion to claim an exceptional treatment, most of all for Christianity. Christianity enjoyed no privileges and claimed no immunities when it boldly confronted and confounded the most ancient and the most powerful religions of the world. Even at present it craves no mercy, and it receives no mercy from those whom our missionaries have to meet face to face in every part of the world; and unless our religion has ceased to be what it was, its defenders should not shrink from this new trial of strength, but should encourage rather than depreciate the study of comparative theology.

And let me remark this, in the very beginning, that no other religion, with the exception, perhaps, of early Buddhism, would have favored the idea of an impartial comparison of the principal religions of the world — would have tolerated our science. Nearly every religion seems to adopt the language of the Pharisee rather than of the publican. It is Christianity alone which, as the religion of humanity, as the religion of no caste, of no chosen people, has taught us to respect the history of humanity, as a whole, to discover the traces of a divine wisdom and love in the government of all the races of mankind, and to recognize, if possible, even in the lowest and crudest forms of religious belief, not the work of demoniacal agencies, but something that indicates a divine guidance, something that makes us perceive, with St. Peter, " that God is no respecter of persons, but that in every nation he that feareth Him and worketh righteousness is accepted with Him."

In no religion was there a soil so well prepared for the cultivation of Comparative Theology as in our own. The position which Christianity from the very beginning took up with regard to Judaism, served as the first lesson in comparative theology, and directed the attention, even of the unlearned, to a comparison of two religions, differing in their conception of the Deity, in their estimate of humanity, in their motives of morality, and in their hope of immortality, yet sharing so much in common that there are but few of the psalms and prayers in the Old Testament in which a Christian cannot heartily join even now, and but few rules of morality which he ought not even now to obey. If we have once learned to see in the exclusive religion of the Jews a preparation of what was to be the all-embracing religion of humanity, we shall feel much less difficulty in recognizing in the mazes of other religions a hidden purpose; a wandering in the desert, it may be, but a preparation also for the land of promise.

A study of these two religions, the Jewish and the Christian, such as it has long been carried on by some of our most learned divines, simultaneously with the study of Greek and Roman mythology, has, in fact, served as a most useful preparation for wider inquiries. Even the mistakes that have been committed by earlier scholars have proved useful to those who followed after; and, once corrected, they are not likely to be committed again. The opinion, for instance, that the pagan religions were mere corruptions of the religion of the Old Testament, once supported by men of high authority and great learning, is now as completely surrendered as the attempts of explaining Greek and Latin as corruptions of Hebrew. The theory again, that there was a primeval preternatural revelation granted to the fathers of the human race, and that the grains of truth which catch our eye when exploring the temples of heathen idols, are the scattered fragments of that sacred heirloom, — the seeds that fell by the way-side or upon stony places, — would find but few supporters at present; no more, in fact, than the theory that there was in the beginning one complete and perfect primeval language, broken up in later times into the numberless languages of the world.

Some other principles, too, have been established within this limited sphere by a comparison of Judaism and Christianity with the religions of Greece and Rome, which will prove extremely useful in guiding us in our own researches. It has been proved, for instance, that the language of antiquity is not like the language of our own times; that the language of the East is not like the language of the West; and that, unless we make allowance for this, we cannot but misinterpret the utterances of the most ancient teachers and poets of the human race. The same words do not mean the same thing in Anglo-Saxon and English, in Latin and French: much less can we expect that the words of any modern language should be the exact equivalents of an ancient Semitic language, such as the Hebrew of the Old Testament.

Ancient words and ancient thoughts, for both go together, have not yet arrived at that stage of abstraction in which, for instance, active powers, whether natural or supernatural, can be represented in any but a personal and more or less human form. When we speak of a temptation from within or from without, it was more natural for the ancients to speak of a tempter, whether in a human or in an animal form; when we speak of the ever-present help of God, they call the Lord their rock, and their fortress, their buckler, and their high tower; what with us is a heavenly message, or a godsend, was to them a winged messenger; what we call divine guidance, they speak of as a pillar of a cloud to lead them the way, and a pillar of light to give them light; a refuge from the storm, and a shadow from the heat. What is really meant is no doubt the same, and the fault is ours, not theirs, if we willfully misinterpret the language of ancient prophets, if we persist in understanding their words in their outward and material aspect only, and forget that before language had sanctioned a distinction between the concrete and the abstract, between the purely spiritual as opposed to the coarsely material, the intention of the speakers comprehends both the concrete and the abstract, both the material and the spiritual, in a manner which has become quite strange to us, though it lives on in the language of every true poet. Unless we make allowance for this mental parallax, all our readings in the ancient skies will be, and must be erroneous. Nay, I believe it can be proved that more than half of the difficulties in the history of religious thought owe their origin to this constant misinterpretion of ancient language by modern language, of ancient thought by modern thought.

That much of what seems to us, and seemed to the best among the ancients, irrational and irreverent in the mythologies of India, Greece, and Italy, can thus be removed, and that many of their childish fables can thus be read again in their original child-like sense, has been proved by the researches of Comparative Mythologists. The phase of language which gives rise, inevitably, we may say, to these misunderstandings, is earlier than the earliest literary documents. Its work in the Aryan languages was done before the time of the Veda, before the time of Homer, though its influence continues to be felt to a much later period.

Is it likely that the Semitic languages, and, more particularly, Hebrew, should, as by a miracle, have escaped the influence of a process which is inherent in the very nature and growth of language, which, in fact, may rightly be called an infantine disease, against which no precautions can be of any avail?

And if it is not, are we likely to lose anything if we try to get at the most ancient, the most original intention of sacred traditions, instead of being satisfied with their later aspect, their modern misinterpretations? Have we lost anything if, while reading the story of Hephæstos splitting open with his axe the head of Zeus, and Athene springing from it full armed, we perceive behind this savage imagery, Zeus as the bright Sky, his forehead as the East, Hephæstos as the young, not yet risen Sun, and Athene as the Dawn, the daughter of the Sky, stepping forth from the fountain-head of light —

Γλαυκῶπις, with eyes like an owl (and beautiful they are);

Παρθένος, pure as a virgin;

Χρύσεα, the golden;

Ἀκρία, lighting up the tops of the mountains, and her own glorious Parthenon in her own favorite town of Athens;

Παλλάς, whirling the shafts of light;

'Αλέα, the genial warmth of the morning;

Πρόμαχος, the foremost champion in the battle between night and day;

Πάνοπλος, in full armor, in her panoply of light, driving away the darkness of night, and awakening men to a bright life, to bright thoughts, to bright endeavors.

Would the Greeks have had less reverence for their gods if, instead of believing that Apollo and Artemis murdered the twelve children of Niobe, they had perceived that Niobe was, in a former period of language, a name of snow and winter, and that no more was intended by the ancient poet than that Apollo and Artemis, the vernal deities, must slay every year with their darts the brilliant and beautiful but doomed children of the Snow? Is it not something worth knowing, worth knowing even to us after the lapse of four or five thousand years, that before the separation of the Aryan race, before the existence of Sanskrit, Greek, or Latin, before the gods of the Veda had been worshipped, and before there was a sanctuary of Zeus among the sacred oaks of Dodona, one supreme deity had been found, had been named, had been invoked by the ancestors of our race, and had been invoked by a name which has never been excelled by any other name?

No; if a critical examination of the ancient language of the Jews leads to no worse results than those which have followed from a careful interpretation of the petrified language of ancient India and Greece, we need not fear; we shall be gainers, not losers. Like an old precious medal, the ancient religion, after the rust of ages has been removed, will come out in all its purity and brightness; and the image which it discloses will be the image of the Father, the Father of all the nations upon earth; and the superscription, when we can read it again, will be, not only in Judæa, but in the languages of all the races of the world, the Word of God, revealed, where alone it can be revealed, — revealed in the heart of man.

SECOND LECTURE.

THERE is no lack of materials, and there is abundance of work for the student of the Science of Religion. It is true that, compared with the number of languages which the comparative philologist has to deal with, the number of religions is small. In a comparative study of languages, however, we find most of our materials ready for use; we possess grammars and dictionaries. But where are we to look for the grammars and dictionaries of the principal religions of the world? Not in the catechisms or the articles, not even in the so-called creeds or confessions of faith which, if they do not give us an actual misrepresentation of the doctrines which they profess to epitomize, give us always the shadow only, and never the soul and substance of a religion. But how seldom do we find even such helps!

Among Eastern nations it is not unusual to distinguish between religions that are founded on a book, and others that have no such vouchers to produce. The former are considered more respectable, and, though they may contain false doctrine, they are looked upon as a kind of aristocracy among the vulgar and nondescript crowd of bookless or illiterate religions.

To the student of religion canonical books are, no doubt, of the utmost importance, though he ought never to forget that nearly all canonical books give the reflected image only of the real doctrines of the founder of a new religion, an image always blurred and distorted by the medium through which it had to pass. But how few are the religions which possess even a sacred canon, how small is the aristocracy of real book-religions in the history of the world! Let us look at the two families that have been the principal actors in that great drama which we call the history of the world, the *Aryan* and the *Semitic,* and we shall find that two members only of each family can claim the possession of a sacred code. Among the *Aryans,* the *Hindus* and the *Persians;* among the *Shemites,* the *Hebrews* and the *Arabs.* In the Aryan family the Hindus, in the Semitic family the Hebrews, have each produced two book-religions; the Hindus have given rise to Brahmanism and Buddhism; the Hebrews to Mosaism and Christianity. Nay, it is important to observe that in each family the third book-religion can hardly lay claim to an independent origin, but is only a weaker repetition of the first. Zoroastrianism has its sources in the same stratum which fed the deeper and broader stream of Vedic religion; Mohammedanism springs, as far as its most vital doctrines are concerned, from the ancient fountain-head of the religion of Abraham, the worshipper and the friend of the one true God. If you keep before your mind the foregoing simple outline, you can see the river system in which the religious thought of the Aryan and the Semitic nations has been running for centuries, — of those, at least, who are in possession of sacred and canonical books.

While Buddhism is the direct offspring, and, at the same time the antagonist of Brahmanism, Zoroastrianism is rather a deviation from the straight course of ancient Vedic faith, though it likewise contains a protest against some of the doctrines of the earliest worshippers of the Vedic gods. The same, or nearly the same relationship holds together the three principal religions of the Semitic stock, only that, chronologically, Mohammedanism is later than Christianity, while Zoroastrianism is earlier than Buddhism.

Observe also another, and, as we shall see, by no means accidental coincidence in the parallel ramifications of these two religious stems.

Buddhism, which is the offspring of, but at the same time marks a reaction against the ancient Brahmanism of India, withered away after a time on the soil from which it had sprung, and assumed its real importance only after it had been transplanted from India, and struck root among Turanian nations in the very centre of the Asiatic continent. Buddhism, being at its birth an Aryan religion, ended by becoming the principal religion of the Turanian world.

The same transference took place in the second stem. Christianity, being the offspring of Mosaism, was rejected by the Jews as Buddhism was by the Brahmans. It failed to fullfil its purpose as a mere reform of the ancient Jewish religion, and not till it had been transferred from Semitic to Aryan ground, from the Jews to the Gentiles, did it develop its real nature and assume its world-wide importance. Having been at its birth a Semitic religion, it became the principal religion of the Aryan world.

There is one other nation only, outside the pale of the Aryan and Semitic families, which can claim one, or even two book-religions as its own. China became the mother, at almost the same time, of two religions, each founded on a sacred code, — the religion of Confucius, and the religion of Lao-tse, the former resting on the Five King and the Four Shu, and the latter on the Tao-te-king.

With these eight religions the library of the Sacred Books of the whole human race is complete, and an accurate study of these eight codes, written in Sanskrit, Pâli, and Zend, in Hebrew, Greek, and Arabic, lastly in Chinese, might in itself not seem too formidable an undertaking for a single scholar. Yet, let us begin at home, and look at the enormous literature devoted to the interpretation of the Old Testament, at the number of books published every year on controverted points in the doctrine or the history of the Gospels, and you may then form an idea of what a theological library would be that should contain the necessary materials for an accurate and scholarlike interpretation of the eight sacred codes. Even in so modern, and, in the beginning, at least, so illiterate a religion as that of Mohammed, the sources that have to be consulted for the history of the faith during the early centuries of its growth are so abundant, that few critical scholars could master them in their completeness.[1]

If we turn our eyes to the Aryan religions, the sacred writings of the Brahmans, in the narrowest acceptation of the word, might seem within easy grasp. The hymns of the Rig-veda, which are the real bible of the ancient faith of the Vedic Rishis, are only 1,028

[1] Sprenger, *Das Leben des Mohammed*, vol. i. p. 9. "Die Quellen, die ich benutzt habe, sind so zahlreich, und der Zustand der Gelehrsamkeit war unter den Moslimen in ihrer Urzeit von dem upsrigen so verschieden, dass die Materialien, die ich uber die Quellen gesammelt habe, ein ziemlich beleibtes Bandchen bilden werden. Es ist in der That nothwendig, die Literaturgeschichte des Islam der ersten zwei Jahrhunderte zu schreiben, um den Leser in den Stand zu setzen, den hier gesammelten kritischen Apparat zu benutzen. Ich gedenke die Resultate meiner Forschungen als ein separates Werkchen nach der Prophetenbiographie herauszugeben."

in number, consisting of about 10,580 verses.[1] The commentary, however, on these hymns, of which I have published four good sized quarto volumes, is estimated at 100,000 lines, consisting of thirty-two syllables each, that is at 3,200,000 syllables. There are besides, the three minor Vedas, the Yagur-veda, the Sâma-veda, the Artharva-veda, which, though of less importance for religious doctrines, are indispensable for a right appreciation of the ceremonial system of the worshippers of the ancient Vedic gods.

To each of these four Vedas belong collections of so-called *Brahmanas*, scholastic treatises of a later time, it is true, but nevertheless written in archaic Sanskrit, and reckoned by every orthodox Hindu as part of his revealed literature. Their bulk is much larger than that of the ancient Vedic hymn-books.

And all this constitutes the text only for numberless treatises, essays, manuals, glosses, etc., forming an uninterrupted chain of theological literature, extending over more than three thousand years, and receiving new links even at the present time. There are, besides, the inevitable parasites of theological literature, the controversial writings of different schools of thought and faith, all claiming to be orthodox, yet differing from each other like day and night; and lastly, the compositions of writers, professedly unorthodox, professedly at variance with the opinions of the majority, declared enemies of the Brahmanic faith and the Brahmanic priesthood, whose accusations and insinuations, whose sledge-hammers of argument, and whose poisoned arrows of invective need fear no comparison with the weapons of theological warfare in any other country.

[1] Max Müller, *History of Ancient Sanskrit Literature*, p. 220.

Nor can we exclude the sacred law books, nor the ancient epic poems, the Mahabharata and Ramayana, nor the more modern, yet sacred literature of India, the Puranas and Tantras, if we wish to gain an insight into the religious belief of millions of human beings, who though they all acknowledge the Veda as their supreme authority in matters of faith, yet are unable to understand one single line of it, and in their daily life depend entirely for spiritual food on the teaching conveyed to them by these more recent and more popular books. And even then our eye would not have reached many of the sacred recesses in which the Hindu mind has taken refuge, either to meditate on the great problem of life, or to free itself from the temptations and fetters of worldly existence by penances and mortifications of the most exquisite cruelty. India has always been teeming with religious sects, and as far as we can look back into the history of that marvelous country, its religious life has been broken up into countless local centres which it required all the ingenuity and perseverance of a priestly caste to hold together with a semblance of dogmatic uniformity. Some of these sects may almost claim the title of independent religions, as, for instance, the once famous sect of the Sikhs, possessing their own sacred code and their own priesthood, and threatening for a time to become a formidable rival of Brahmanism and Mohammedanism in India. Political circumstances gave to the sect of Nanak its historical prominence and more lasting fame. To the student of religion it is but one out of many sects which took their origin in the fifteenth and sixteenth centuries, and attempted to replace the corruptions of Hinduism and Mohammedanism by a purer and more spiritual worship. The Granth, *i. e.* the

Volume, the sacred book of the Sikhs, is full of interest, full of really deep and poetical thought; and it is to be hoped that it will soon find an English translator. But there are other collections of religious poetry, more ancient and more original than the stanzas of Nanak; nay, many of the most beautiful verses of the Granth were borrowed from these earlier authorities, particularly from Kabir, the pupil of Ramanand. Here there is enough to occupy the students of religion: an intellectual flora of greater variety and profuseness than even the natural flora of that fertile country.

And yet we have not said a word as yet of the second book-religion of India — of the religion of Buddha, originally one only out of numberless sects, but possessing a vitality which has made its branches to overshadow the largest portion of the inhabited globe. Who can say — I do not speak of European scholars only, but of the most learned members of the Buddhist fraternities — who can say that he has read the whole of the canonical books of the Buddhist Church, to say nothing of their commentaries or later treatises? The text and commentaries of the Buddhist canon contain, according to a statement in the Saddharma-alankara,[1] 29,368,000 letters. Such statements do not convey to our mind any very definite idea, nor could any scholar vouch for their absolute correctness. But if we consider that the English Bible is said to contain about three millions and a half of letters,[2] (and here vowels are counted separately from consonants), five or six times that amount would hardly seem enough, as a rough estimate of the bulk of the Buddhist scriptures. The Tibetan edition of the Buddhist canon, consisting of

---

[1] Spence Hardy, *The Legends and Theories of the Buddhists*, p. 66.
[2] 3,567,180.

two collections, the Kanjur and Tanjur, numbers, about three hundred and twenty-five volumes folio, each weighing, in the Pekin edition, from four to five pounds.[1]

Apparently within a smaller compass lies the sacred literature of the third of the Aryan book-religions, the so-called Zendavesta. But here the very scantiness of the ancient text increases the difficulty of its successful interpretation, and the absence of native commentaries has thrown nearly the whole burden of deciphering on the patience and ingenuity of European scholars.

If lastly we turn to China, we find that the religion of Confucius is founded on the Five King and the Four Shu — books in themselves of considerable extent, and surrounded by voluminous commentaries, without which even the most learned scholars would not venture to fathom the depth of their sacred canon.[2]

Lao-tse, the contemporary or rather the senior of Confucius, is reported to have written a large number of books;[3] no less than nine hundred and thirty on different questions of faith, morality, and worship, and seventy on magic. His principal work, however, the Tao-te-king, which represents the real scripture of his followers, the Tao-sse, consists only of about five thousand words,[4] and fills no more than thirty pages. But here again we find that for that very reason the text is unintelligible without copious commentaries, so that M. Julien had to consult more than sixty commentators for

---

[1] *Chips from a German Workshop*, vol. i. p. 193.

[2] *The Chinese Classics*, with a Translation, Notes, Prolegomena, and Indexes. By James Legge, D. D.; 7 vols. London: Trübner & Co.

[3] Stan. Julien, *Tao te king*, p. xxvii.

[4] Julien, *Tao te king*, p. xxxi., xxxv. The texts vary from 5,610, 5,630, 5,688, to 5,722 words. The text published by M. Stan. Julien consists of 5,320 words.

the purpose of his translation, the earliest going back as far as the year 163 B. C.

There is a third established religion in China, that of Fo; but Fo is only the Chinese corruption of Buddha, and though the religion of Buddha, as transferred from India to China, has assumed a peculiar character and produced an enormous literature of its own, yet Chinese Buddhism cannot be called an independent religion, any more than Buddhism in Ceylon, Burmah, and Siam, or in Nepaul, Tibet, and Mongolia.

But after we have collected this library of the sacred books of the world with their indispensable commentaries, are we then in possession of the requisite materials for studying the growth and decay of the religious convictions of mankind at large? Far from it. The largest portion of mankind,— aye, and some of the most valiant champions in the religious and intellectual struggles of the world, would be unrepresented in our theological library. Think only of the Greeks and the Romans; think of the Teutonic, the Celtic, and Slavonic nations! Where are we to gain an insight into what we may call their real religious convictions, previous to the comparatively recent period when their ancient temples were leveled to the ground to make room for new cathedrals; and their sacred oaks were felled to be changed into crosses, planted along every mountain pass and forest lane? Homer and Hesiod do not tell us what was the religion, the real heart-religion of the Greeks, nor were their own poems ever considered as sacred, or even as authoritative and binding, by the highest intellects among the Greeks. In Rome we have not even an Iliad or Odyssey; and when we ask for the religious worship of the Teutonic, the Celtic, or the Slavonic tribes, the very names of many of the deities in whom they believed are forgotten and lost forever, and the scattered notices of their faith have to be picked up and put together like the small stones of a broken mosaic that once formed the pavement in the ruined temples of Rome.

The same gaps, the same want of representative authorities, which we witness among the Aryan, we meet again among the Semitic nations, as soon as we step out of the circle of their book-religions. The Babylonians, the Phenicians, and Carthaginians, the Arabs before their conversion to Mohammedanism, all are without canonical books, and a knowledge of their religion has to be gathered, as well as may be, from monuments, inscriptions, traditions, from proper names, from proverbs, from curses, and other stray notices which require the greatest care before they can be properly sifted and successfully fitted together.

But now let us go on further. The two beds in which the stream of Aryan and Semitic thought has been rolling on for centuries from southeast to northwest, from the Indus to the Thames, from the Euphrates to the Jordan and the Mediterranean, cover but a narrow tract of country compared with the vastness of our globe. As we rise higher, our horizon expands on every side, and wherever there are traces of human life there are traces also of religion. Along the shores of the ancient Nile we see still standing the Pyramids, and the ruins of temples and labyrinths, their walls covered with hieroglyphic inscriptions, and with the strange pictures of gods and goddesses. On rolls of papyrus, which seem to defy the ravages of time, we have even fragments of what may be called the sacred books of the Egyptians. Yet though much has been deciphered in the ancient records of that mysterious race, the mainspring of the religion of Egypt and the original intention of its ceremonial worship are far from being fully disclosed to us. As we follow the sacred stream to its distant sources the whole continent of Africa opens before us, and wherever we now see kraals and cattle-pens, depend upon it there was to be seen once, or there is to be seen even now, the smoke of sacrifices rising up from earth to heaven. The ancient relics of African faith are rapidly disappearing at the approach of Mohammedan and Christian missionaries; but what has been preserved of it, chiefly through the exertions of learned missionaries, is full of interest to the student of religion, with its strange worship of snakes and ancestors, its vague hope of a future life, and its not altogether faded reminiscence of a Supreme God, the Father of the black as well as of the white man.

From the eastern coast of Africa our eye is carried across the sea where, from Madagascar to Hawaii, island after island stands out like so many pillars of a sunken bridge that once spanned the Indian and Pacific oceans. Everywhere, whether among the dark Papuan or the yellowish Malay, or the brown Polynesian races scattered on these islands, even among the lowest of the low in the scale of humanity, there are, if we will but listen, whisperings about divine beings, imaginings of a future life; there are prayers and sacrifices which, even in their most degraded and degrading form, still bear witness to that old and ineradicable faith that everywhere there is a God to hear our prayers, if we will but call on Him, and to accept our offerings, if they are offered as a ransom for sin or as a token of a grateful heart.

Still farther east the double continent of America

becomes visible, and in spite of the unchristian vandalism of its first discoverers and conquerors, there, too, we find materials for the study of an ancient, and, it would seem, independent faith. Unfortunately, the religious and mythological traditions, collected by the first Europeans who came in contact with the natives of America, reach back but a short distance beyond the time when they were written down, and they seem in several cases to reflect the thoughts of the Spanish listeners as much as those of the native narrators. The quaint hieroglyphic manuscripts of Mexico and Guatemala have as yet told us very little, and the accounts written by natives in their native language have to be used with great caution. Still the ancient religion of the Aztecs of Mexico and of the Incas of Peru is full of interesting problems. As we advance towards the north and its red skinned inhabitants, our information becomes more meagre still, and after what happened some years ago, no " Livre des Sauvages " is likely to come to our assistance again. Yet there are wild and home-grown specimens of religious faith to be studied even now among the receding and gradually perishing tribes of the Red Indians, and, in their languages as well as in their religions, traces may possibly still be found, before it is too late, of pre-historic migrations of men from the primitive Asiatic to the American continent, either across the stepping-stones of the Aleutic bridge in the north, or lower south by drifting with favorable winds from island to island, till the hardy canoe was landed or wrecked on the American coast, never more to return to the Asiatic home from which it started.

And when in our religious survey we finally come back again to the Asiatic continent, we find here too, although nearly the whole of its area is now occupied by one or the other of the eight book-religions, by Mosaism, Christianity, and Mohammedanism, by Brahmanism, Buddhism, and Zoroastrianism, and in China by the religions of Confucius and Lao-tse, that nevertheless partly below the surface, and in some places still on the surface, more primitive forms of worship have maintained themselves. I mean the Shamanism of the Mongolian race, and the beautiful half-Homeric mythology of the Finnish and Esthonian tribes.

And now that I have displayed this world-wide panorama before your eyes, you will share, I think, the feeling of dismay with which the student of the science of religion looks around and asks himself where to begin and how to proceed. That there are materials in abundance, capable of scientific treatment, no one would venture to deny. But how are they to be held together? How are we to discover what all these religions share in common? How they differ? How they rise and how they decline? What they are and what they mean?

Let us take the old saying, *Divide et impera*, and translate it somewhat freely by " Classify and conquer," and I believe we shall then lay hold of the old thread of Ariadne which has led the students of many a science through darker labyrinths even than the labyrinth of the religions of the world. All real science rests on classification, and only in case we cannot succeed in classifying the various dialects of faith shall we have to confess that a science of religion is really an impossibility. If the ground before us has once been properly surveyed and carefully parceled out, each scholar may then cultivate his own glebe, without wasting his energies and without losing sight of the general purposes to which all special researches must be subservient.

How, then, is the vast domain of religion to be parceled out? How are religions to be classified, or, we ought rather to ask first, how have they been classified before now? The simplest classification, and one which we find adopted in almost every country, is that into *true* and *false* religions. It is very much like the first classification of languages into one's own language and the language of the rest of the world; as the Greeks would say, into the languages of the Greeks and the Barbarians; or, as the Jews would say, into the languages of the Jews and the Gentiles; or, as the Hindus would say, into the languages of the Aryas and Mle*kkh*as; or, as the Chinese would say, into the languages of the Middle Empire and that of the Outer Barbarians. I need not say why that sort of classification is useless for scientific purposes.

There is another classification, apparently of a more scientific character, but if examined more closely, equally worthless to the student of religion. I mean the well-known division into *revealed* and *natural* religions.

I have first to say a few words on the meaning attached to Natural Religion. That word is constantly used in very different acceptations. It is applied by several writers to certain historical forms of religion, which are looked upon as not resting on the authority of revelation, in whatever sense that word may be hereafter interpreted. Thus Buddhism would be a natural religion in the eyes of the Brahmans, Brahmanism would be a natural religion in the eyes of the Mohammedans. With us, all religions except Christianity, and, though in a lesser degree, Mosaism, would be classed as merely natural; and though natural does not imply false, yet it distinctly implies the absence of any sanction beyond the sense of truth, or the voice of conscience that is within us.

But Natural Religion is also used in a very different sense, particularly by the philosophers of the last century. When people began to subject the principal historical religions to a critical analysis, they found that after removing what was peculiar to each, there remained certain principles which they all shared in common. These were supposed to be the principles of Natural Religion. Again, when everything that seemed supernatural, miraculous, and irrational, had been removed from the pages of the New Testament, there still remained a kind of skeleton of religion, and this too was passed off under the name of Natural Religion. During the last century, philosophers who were opposing the spread of skepticism and infidelity, thought that this kind of natural, or, as it was also called, rational religion, might serve as a breakwater against utter unbelief; but they soon found out that a mere philosophical system, however true, can never take the place of religious faith. When Diderot said that all revealed religions were the heresies of Natural Religion, he meant by Natural Religion a body of truths implanted in human nature, to be discovered by the eye of reason alone, and independent of any such historical or local influences as give to each religion its peculiar character and local aspect. The existence of a deity, the nature of his attributes, such as Omnipotence, Omniscience, Omnipresence, Eternity, Self-existence, Spirituality, the Goodness also of the Deity, and connected with it, the admission of a distinction between Good and Evil, between Virtue and Vice, all this, and according to some writers, the Unity and Personality also of the Deity, were included in the domain of Natural Religion. The scientific treatment of this so-called Natural Religion received the name of Natural Theology, a title rendered famous in the beginning of our century by the much praised and much abused work of Paley. Natural Religion corresponds in the science of religion to what in the science of language used to be called *Grammaire générale*, a collection of fundamental rules which are supposed to be self-evident, without which no grammar would be possible, but which, strange to say, never exist in their purity and completeness in any language that is or ever has been spoken by human beings. It is the same with religion. There never has been any real religion, consisting exclusively of the pure and simple tenets of Natural Religion, though there have been certain philosophers who brought themselves to believe that their religion was entirely rational, — was, in fact, pure and simple Deism.

If we speak, therefore, of a classification of all historical religions into revealed and natural, what is meant by natural is simply the negation of revealed, and if we tried to carry out the classification practically, we should find the same result as before. We should have on one side Christianity alone, or, according to some theologians, Christianity and Judaism; on the other, all the remaining religions of the world.

This classification, therefore, whatever may be its practical value, is perfectly useless for scientific purposes. A more extended study shows us very soon that the claim of revelation is set up by the founders, or if not by them, at all events by the later preachers and advocates of most religions; and would therefore be declined by all but ourselves as a distinguishing feature of Christianity and Judaism. We shall see, in fact, that the claims to a revealed authority are urged far more strongly and elaborately by the believers in the Veda, than by the apologetical theologians among the Jews and Christians. Even Buddha, originally the most thoroughly human and self-dependent among the founders of religion, is by a strange kind of inconsistency represented in later controversial writings, as in possession of revealed truth.[1] He himself could not, like Numa, or Zoroaster, or Mohammed,[2] claim communication with higher spirits; still less could he, like the poets of the Veda, speak of divine inspirations and god-given utterances: for according to him there was none among the spirits greater or wiser than himself, and the gods of the Veda had become his servants and worshippers. Buddha himself appeals only to what we should call the inner light.[3] When he delivered for the first time the four fundamental doctrines of his system, he said, "Mendicants, for the attainment of these previously unknown doctrines, the eye, the knowledge, the wisdom, the clear perception, the light were developed within me." He was called Sarvagna or omniscient by his earliest pupils; but when in later times it was seen that on several points Buddha had but spoken the language of his age, and had shared the errors current among his contemporaries with regard to the shape of the earth and the movement of the heavenly bodies, an important concession was made by Buddhist theologians. They

---

[1] *History of Ancient Sanskrit Literature,* by Max Müller, p. 83.
[2] Sprenger, *Mohammed,* vol. ii. p. 426.
[3] Gogerly, *The Evidences and Doctrines of Christian Religion.* Colombo, 1862. Part i.

limited the meaning of the word "omniscient," as applied to Buddha, to a knowledge of the principal doctrines of his system, and concerning these, but these only, they declared him to have been infallible. This may seem to be a modern kind of view, but whether modern or ancient, it certainly reflects great credit on the Buddhist theologians. In the Milinda Prasna, however, which is a canonical book, we see that the same idea was already rising in the mind of the great Nagasena. Being asked by King Milinda whether Buddha is omniscient, he replies: "Yes, Great King, the blessed Buddha is omniscient. But Buddha does not at all times exercise his omniscience. By meditation he knows all things; meditating he knows everything he desires to know." In this reply a distinction is evidently intended between subjects that may be known by sense and reason, and subjects that can be known by meditation only. Within the domain of sense and reason, Nagasena does not claim omniscience or infallibility for Buddha, but he claims for him both omniscience and infallibility in all that is to be perceived by meditation only, or, as we should say, in matters of faith.

I shall have to explain to you hereafter the extraordinary contrivances by which the Brahmans endeavored to eliminate every human element from the hymns of the Veda, and to establish, not only the revealed, but the prehistoric or even antemundane character of their scriptures. No apologetic writings have ever carried the theory of revelation to greater extremes.

In the present stage of our inquiries, all that I wish to point out is this, — that when the founders or defenders of nearly all the religions of the world appeal to some kind of revelation in support of the truth of their doctrines, it could answer no useful purpose were we to attempt any classification on such disputed ground. Whether the claim of a natural or preternatural revelation, put forward by different religions, is well founded or not, is not the question at present. It falls to the province of Theoretic Theology to explain the true meaning of revelation, for few words have been used so vaguely and in so many different senses. It falls to its province to explain, not only how the veil was withdrawn that intercepted for a time the rays of divine truth, but what is a far more difficult problem, how there could ever have been a veil between truth and the seeker of truth, between the adoring heart and the object of the highest adoration, between the Father and his children.

In Comparative Theology our task is different: we have simply to deal with the facts such as we find them. If people regard their religion as revealed, it is to them a revealed religion, and has to be treated as such by an impartial historian. We cannot determine a question by adopting, without discussion, the claims of one party, and ignoring those of the other.

But this principle of classification into revealed and natural religions appears still more faulty, when we look at it from another point of view. Even if we granted that all religions, except Christianity and Mosaism, derived their origin from those faculties of the mind only which, according to Paley, are sufficient by themselves for calling into life the fundamental tenets of what we explained before as natural religion, the classification of Christianity and Judaism on one side as *revealed*, and of the other religions as *natural*, would still be defective, for the simple reason that no religion, though founded on revelation, can ever be entirely separated from natural religion. The tenets of natural religion, though by themselves they never constituted a real historical religion, supply the only ground on which revealed religion can stand, the only soil where it can strike root, and from which it can receive nourishment and life. If we took away that soil, or if we supposed that it, too, had to be supplied by revelation, we should not only run counter to the letter and spirit of the Old and the New Testament, but we should degrade revealed religion by changing it into a mere formula, to be accepted by a recipient incapable of questioning, weighing, and appreciating its truth; we should indeed have the germ, but we should have thrown away the congenial soil in which alone that germ of true religion can live and grow.

Christianity, addressing itself not only to the Jews, but also to the Gentiles, not only to the ignorant, but also to the learned, not only to the believers, but in the first instance, to the unbeliever, presupposed in all of them the elements of natural religion, and with them the power of choosing between truth and untruth. Thus only could St. Paul say: "Prove all things; hold fast that which is good." (1 Thess. v. 21.)

The same is true with regard to the Old Testament. There, too, the belief in a Deity, and in some at least of its indefeasible attributes, is taken for granted, and the prophets who call the wayward Jews back to the worship of Jehovah, appeal to them as competent, by the truth-testing power that is within them, to choose between Jehovah and the gods of the Gentiles, between truth and untruth. Remember only the important chapter in the earliest history of the Jews, when Joshua gathered all the tribes of Israel to Shechem, and called for the elders of Israel, and for their heads, and for their judges, and for their officers; and they presented themselves before God.

"And Joshua said unto all the people: Thus saith the Lord God of Israel: Your fathers dwelt on the other side of the flood in old time, even Terah, the father of Abraham, and the father of Nachor; and they served other gods."

And then, after reminding them of all that God has done for them, he concludes by saying:—

"Now, therefore, fear the Lord, and serve Him in sincerity and in truth; and put away the gods which your fathers served on the other side of the flood, and in Egypt, and serve ye the Lord.

"And if it seem evil unto you to serve the Lord, *choose you* this day whom ye will serve; whether the gods which your fathers served that were on the other side of the flood, or the gods of the Amorites in whose lands ye dwell; but as for me and my house, we will serve the Lord."

In order to choose between different gods and different forms of faith, a man must possess the faculty of choosing, the instruments of testing truth and untruth, whether revealed or not; he must know that certain fundamental tenets cannot be absent in any true religion, and that there are doctrines against which his rational or moral conscience revolts as incompatible with truth. In short, there must be the foundation of religion, there must be the solid rock, before it is possible to erect an altar, a temple, or a church; and if we call that foundation natural religion, it is clear that no revealed religion can be thought of which does not rest more or less firmly on natural religion.

These difficulties have been felt distinctly by some

of our most learned divines, who have attempted a classification of religions from their own point of view. New definitions of natural religion have therefore been proposed in order to avoid the overlapping of the two definitions of natural and revealed religion. Natural religion has, for instance, been explained as the religion of nature before revelation, such as may be supposed to have existed among the patriarchs, or to exist still among primitive people who have not yet been enlightened by Christianity or debased by idolatry.

According to this view we should have to distinguish, not two, but three classes of religion: the primitive or natural, the debased or idolatrous, and the revealed. But, as pointed out before, the first, the so-called primitive or natural religion, exists in the minds of modern philosophers rather than of ancient poets and prophets. History never tells us of any race with whom the simple feeling of reverence for higher powers was not hidden under mythological disguises. Nor would it be possible even thus to separate the three classes of religion by sharp and definite lines of demarcation, because both the debased or idolatrous and the purified or revealed religions would, of necessity, include within themselves the elements of natural religion. Nor do we diminish these difficulties in the classificatory stage of our science, if, in the place of this simple natural religion, we admit with other theologians and philosophers, a universal primeval revelation. This universal primeval revelation is only another name for natural religion, and it rests on no authority but the speculations of philosophers. The same class of philosophers, considering that language was too wonderful an achievement for the human mind, insisted on the necessity of admitting a universal primeval language revealed directly by God to man, or rather to mute beings; while the more thoughtful and the more reverent among the Fathers of the Church and among the founders of modern philosophy pointed out that it was more consonant with the general working of an all-wise and all-powerful Creator, that he should have endowed human nature with germinant faculties of speech, instead of presenting mute beings with grammars and dictionaries ready made. Is an infant less wonderful than a man? an acorn less wonderful than an oak-tree? a cell, if you like, or a protoplasm, including potentially within itself all that it has to become hereafter, less wonderful than all the moving creatures that have life? The same applies in religion. A universal primeval religion revealed direct by God to man, or rather to a crowd of atheists, may, to our human wisdom, seem the best solution of all difficulties; but a higher wisdom speaks to us from out the realities of history, and teaches us, if we will but learn, that " we have all to seek the Lord, if haply we may feel after Him, and find Him, though he be not far from every one of us."

Of the hypothesis of a universal primeval revelation and all its self-created difficulties we shall have to speak again; for the present it must suffice if we have shown that the problem of a scientific classification of religion is not brought nearer to its solution by the additional assumption of another purely hypothetical class of religion.

We have not finished yet. A very important, and, for certain purposes, very useful classification has been that into polytheistic, dualistic, and monotheistic religions. If religion rests chiefly on a belief in a Higher Power, then the nature of that Higher Power would seem to supply a very characteristic feature by which to classify the religions of the world. Nor do I deny that for certain purposes such a classification has proved useful; all I maintain is that we should thus have to class together religions heterogeneous in other respects, though agreeing in the number of their deities. Besides, it would certainly be necessary to add two other classes — the *henotheistic* and the *atheistic*. Henotheistic religions differ from polytheistic because, although they recognize the existence of various deities, or names of deities, they represent each deity as independent of all the rest, as the only deity present in the mind of the worshipper at the time of his worship and prayer. This character is very prominent in the religion of the Vedic poets. Although many gods are invoked in different hymns, sometimes also in the same hymn, yet there is no rule of precedence established among them; and, according to the varying aspects of nature, and the various cravings of the human heart, it is sometimes Indra, the god of the blue sky, sometimes Agni, the god of fire, sometimes Varuṇa, the ancient god of the firmament, who are praised as supreme without any suspicion of rivalry, or any idea of subordination. This peculiar phase of religion, this worship of single gods, forms probably everywhere the first stage in the growth of polytheism, and deserves therefore a separate name.

As to atheistic religions, they might seem to be perfectly impossible; and yet the fact cannot be disputed away that the religion of Buddha was from the beginning purely atheistic. The idea of the Godhead, after it had been degraded by endless mythological absurdities which struck and repelled the heart of Buddha, was, for a time at least, entirely expelled from the sanctuary of the human mind; and the highest morality that was ever taught before the rise of Christianity was taught by men with whom the gods had become mere phantoms, and who had no altars, not even an altar to the Unknown God.

It will be the object of my next lecture to show that the only scientific and truly genetic classification of religions is the same as the classification of languages, and that, particularly in the early history of the human intellect, there exists the most intimate relationship between language, religion, and nationality — a relationship quite independent of those physical elements, the blood, the skull, or the hair, on which ethnologists have attempted to found their classification of the human race.

THIRD LECTURE.

IF we approached the religions of mankind without any prejudices or predilections, in that frame of mind in which the lover of truth or the man of science ought to approach every subject, I believe we should not be long before recognizing the natural lines of demarkation which divide the whole religious world into several great continents. I am speaking, of course, of ancient religions only, or of the earliest period in the history of religious thought. In that primitive period which might be called, if not prehistoric, at least purely ethnic, because what we know of it consists only in the general movements of nations, and not in the acts of individuals, of parties, or of states, — in that primitive period, I say, nations have been called languages; and in our best works on the ancient history of mankind, a map of languages has actually taken the place of a map of nations. But during the same primitive period nations might with equal right be called religions; for there is at that time the same, nay, an even more intimate, relationship between religion and nationality as between language and nationality. In order clearly to explain my meaning, I shall have to refer, as shortly as possible, to the speculations of some German philosophers on the true relation between language, religion, and nationality, — speculations which have as yet received less attention on the part of modern ethnologists than they seem to me to deserve.

It was Schelling, one of the profoundest thinkers of Germany, who first asked the question, What makes an *ethnos?* What is the true origin of a people? How did human beings become a people? And the answer which he gave, though it sounded startling to me when, in 1845, I listened, at Berlin, to the lectures of the old philosopher, has been confirmed more and more by subsequent researches into the history of language and religion.

To say that man is a gregarious animal, and that, like swarms of bees, or herds of wild elephants, men keep together instinctively and thus form themselves into a people, is saying very little. It might explain the agglomeration of one large flock of human beings, but it would never explain the formation of individual peoples.

Nor should we advance much towards a solution of our problem if we were told that men are broken up into peoples as bees are broken up into swarms, by following different queens, by owing allegiance to different governments. Allegiance to the same government, particularly in ancient times, is the result rather than the cause of nationality; while in historical times, such has been the confusion produced by extraneous influences, by brute force, or dynastic combinations, that the natural development of peoples has been entirely arrested, and we frequently find one and the same people divided by different governments, and different peoples united under the same ruler.

Our question, What makes a people? has to be considered in reference to the most ancient times. How did men form themselves into a people before there were kings or shepherds of men! Was it through community of blood? I doubt it. Community of blood produces families, clans, possibly races, but it does not produce that higher and purely moral feeling which binds men together and makes them a people.

It is language and religion that make a people, but religion is even a more powerful agent than language. The languages of many of the aboriginal inhabitants of Northern America are but dialectic varieties of one type, but those who spoke these dialects have never coalesced into a people. They remained mere clans or wandering tribes; they never knew the feeling of a nation because they never knew the feeling of worshipping the same gods. The Greeks, on the contrary, though speaking their strongly marked, and I doubt whether mutually intelligible dialects, the Æolic, the Doric, the Ionic, felt themselves at all times, even when ruled by different tyrants, or broken up into numerous republics, as one great Hellenic people. What was it, then, that preserved in their hearts, in spite of dialects, in spite of dynasties, in spite even of the feuds of tribes and the jealousies of states, the deep feeling of that ideal unity which constitutes a people? It was their primitive religion; it was a dim recollection of the common allegiance they owed from time immemorial to the great father of gods and men; it was their belief in the old Zeus of Dodona, in the Panhellenic Zeus.

Perhaps the most signal confirmation of this view that it is religion even more than language which supplies the foundation of nationality, is to be found in the history of the Jews, the chosen people of God. The language of the Jews differed from that of the Phenicians, the Moabites, and the other neighboring tribes much less than the Greek dialects differed from each other. But the worship of Jehovah made the Jews a peculiar people, the people of Jehovah, separated by their God, though not by their language, from the people of Chemosh (the Moabites)[1] and from the worshippers of Baal and Ashtoreth. It was their faith in Jehovah that changed the wandering tribes of Israel into a nation.

"A people," as Schelling says, "exists only when it has determined itself with regard to its mythology. This mythology, therefore, cannot take its origin after a national separation has taken place, after a people has become a people; nor could it spring up while a people was still contained as an invisible part in the whole of humanity; but its origin must be referred to that very period of transition before a people has assumed its definite existence, and when it is on the point of separating and constituting itself. The same applies to the language of a people; it becomes definite at the same time that a people becomes definite."[2]

Hegel, the great rival of Schelling, arrived at the same conclusion. In his "Philosophy of History" he says: "The idea of God constitutes the general foundation of a people. Whatever is the form of a religion, the same is the form of a state and its constitution; it springs from religion, so much so that the Athenian and the Roman states were possible only with the peculiar heathendom of those peoples, and that even now a Roman Catholic state has a different genius and a different constitution from a Protestant state. The genius of a people is a definite, individ-

---

[1] Numb. xxi. 29; Jer. xlviii. 7: "And Chemosh shall go forth into captivity, with his priests and his princes together."
[2] *Vorlesungen uber Philosophie der Mythologie*, vol i. p. 107, seq.

ual genius, which becomes conscious of its individuality in different spheres; in the character of its moral life, its political constitution, its art, religion and science."[1]

But this is not an idea of philosophers only. Historians, and more particularly the students of the history of law, have arrived at very much the same conclusion. Though to many of them law seems naturally to be the foundation of society, and the bond that binds a nation together, those who look below the surface have quickly perceived that law itself, at least ancient law, derives its authority, its force, its very life from religion. Mr. Maine is no doubt right when, in the case of the so-called Laws of Manu, he rejects the idea of the Deity dictating an entire code or body of law, as an idea of decidedly modern origin. Yet the belief that the lawgiver enjoyed some closer intimacy with the Deity than ordinary mortals, pervades the ancient traditions of many nations. According to a well-known passage in Diodorus Siculus,[2] "the Egyptians believed their laws to have been communicated to Mnevis by Hermes; the Cretans held that Minos received his laws from Zeus, the Lacedæmonians that Lykurgos received his laws from Apollon. According to the Aryans, their lawgiver, Zathraustes, had re-

---

[1] Though these words of Hegel's were published long before Schelling's lectures, they seem to me to breathe the spirit of Schelling rather than of Hegel, and it is but fair therefore to state that Schelling's lectures, though not published, were printed and circulated among friends twenty years before they were delivered at Berlin. The question of priority may seem of little importance on matters such as these, but there is nevertheless much truth in Schelling's remark, that philosophy advances not so much by the answers given to difficult problems, as by the starting of new problems, and by asking questions which no one else would think of asking.

[2] L. i. c. 94.

ceived his laws from the Good Spirit; according to the Getæ, Zamolxis received his laws from the goddess Hestia; and, according to the Jews, Moses received his laws from the god Iao." No one has pointed out more forcibly than Mr. Maine that in ancient times religion as a divine influence was underlying and supporting every relation of life and every social institution. "A supernatural presidency," he writes, " is supposed to consecrate and keep together all the cardinal institutions of those early times, the state, the race, and the family."[1] "The elementary group is the family; the aggregation of families forms the *gens* or the house. The aggregation of houses makes the tribe. The aggregation of tribes constitutes the commonwealth."[2] Now the family is held together by the family *sacra*,[3] and so were the gens, the tribe, and the commonwealth; and strangers could only be admitted to these brotherhoods by being admitted to their *sacra*.[4] At a later time, law breaks away from religion,[5] but even then many traces remain to show that the hearth was the first altar, the father the first elder, his wife and children and slaves the first congregation gathered together round the sacred fire, — the Hestia, the goddess of the house, and in the end the goddess of the people. To the present day marriage, the most important of civil acts, the very foundation of civilized life, has retained the religious character which it had from the very beginning of history.

Let us see now what religion really is in those early ages of which we are here speaking; I do not mean religion as a silent power, working in the heart of man; I mean religion in its outward appearance, religion as something outspoken, tangible, and definite,

[1] Page 6. [2] Page 128. [3] Page 191. [4] Page 131. [5] Page 193.

that can be described and communicated to others. We shall find that in that sense religion lies within a very small compass. A few words, recognized as names of the deity; a few epithets that have been raised from their material meaning to a higher and more spiritual stage, I mean words which expressed originally bodily strength, or brightness, or purity, and which came gradually to mean greatness, goodness, and holiness; lastly, some more or less technical terms expressive of such ideas as *sacrifice*, *altar*, *prayer*, possibly *virtue* and *sin*, *body* and *spirit*, — that is what constitutes the outward framework of the incipient religions of antiquity. If we look at this simple manifestation of religion, we see at once why religion, during those early ages of which we are here speaking, may really and truly be called a sacred dialect of human speech; how at all events early religion and early language are most intimately connected, religion depending entirely for its outward expression on the more or less adequate resources of language.

If this dependence of early religion on language is once clearly understood, it follows, as a matter of course, that whatever classification has been found most useful in the science of language ought to prove equally useful in the science of religion. If there is a truly genetic relationship of languages, the same relationship ought to hold together the religions of the world, at least the most ancient religions.

Before we proceed therefore to consider the proper classification of religions, it will be necessary to say a few words on the present state of our knowledge with regard to the genetic relationship of languages.

If we confine ourselves to the Asiatic continent with its important peninsula of Europe, we find that in the vast desert of drifting human speech three and only three oases have been formed in which, before the beginning of all history, language became permanent and traditional, assumed in fact a new character, a character totally different from the original character of the floating and constantly varying speech of human beings. These three oases of language are known by the name of *Turanian, Aryan,* and *Semitic*. In these three centres, more particularly in the *Aryan* and *Semitic*, language ceased to be natural; its growth was arrested, and it became permanent, solid, petrified, or, if you like, historical speech. I have always maintained that this centralization and traditional conservation of language could only have been the result of religious and political influences, and I now mean to show that we really have clear evidence of three independent settlements of religion, the *Turanian*, the *Aryan*, and the *Semitic*, concomitantly with the three great settlements of language.

Taking Chinese for what it can hardly any longer be doubted that it is, namely, the earliest representative of Turanian speech, we find in China an ancient colorless and unpoetical religion, a religion we might almost venture to call monosyllabic, consisting of the worship of a host of single spirits, representing the sky, the sun, storms and lightning, mountains and rivers, one standing by the side of the other without any mutual attraction, without any higher principle to hold them together. In addition to this, we likewise meet in China with the worship of ancestral spirits, the spirits of the departed, who are supposed to retain some cognizance of human affairs, and to possess peculiar powers which they exercise for good or for evil. This double worship of human and of natural spirits constitutes the old popular religion of China, and it has lived on to the present day, at least in the lower ranks of society, though there towers above it a more elevated range of half religious and half philosophical faith, a belief in two higher Powers, which, in the language of philosophy, may mean *Form* and *Matter*, in the language of Ethics, *Good* and *Evil*, but which in the original language of religion and mythology are represented as *Heaven* and *Earth*.

It is true that we know the ancient popular religion of China from the works of Confucius only, or from even more modern sources. But Confucius, though he is called the founder of a new religion, was really but the new preacher of an old religion. He was emphatically a transmitter, not a maker.[1] He says himself, "I only hand on; I cannot create new things. I believe in the ancients, and therefore I love them."[2]

We find, secondly, the ancient worship of the Semitic races clearly marked by a number of names of the Deity, which appear in the polytheistic religions of the Babylonians, the Phenicians, and Carthaginians, as well as in the monotheistic creeds of Jews, Christians, and Mohammedans. It is almost impossible to characterize the religion of people so different from each other in language, in literature, and general civilization, so different also from themselves at different periods of their history; but if I ventured to characterize the worship of all the Semitic nations by one word, I should say it was preëminently a worship of *God in History*, of God as affecting the destinies of individuals and races and nations rather than of God as wielding the powers of nature. The names of

---

[1] Dr. Legge, *Life of Confucius*, p. 96.
[2] Lun-yu (§ i. a.); Schott, *Chinesische Literatur*, p. 7.

the Semitic deities are mostly words expressive of moral qualities; they mean the Strong, the Exalted, the Lord, the King; and they grow but seldom into divine personalities, definite in their outward appearance, or easily to be recognized by strongly marked features of a real dramatic character. Hence many of the ancient Semitic gods have a tendency to flow together, and a transition from the worship of single gods to the worship of one God required no great effort. In the monotonous desert, more particularly, the worship of single gods glided away almost perceptibly into the worship of one God. If I were to add, as a distinguishing mark, that the Semitic religions excluded the feminine gender in their names of the Deity, or that all their female deities were only representatives of the active energies of older and sexless gods, this would be true of some only, not of all; and it would require nearly as many limitations as the statement of M. Renan, that the Semitic religions were instinctively monotheistic.

We find lastly the ancient worship of the Aryan race, carried to all the corners of the earth by its adventurous sons, and easily recognized, whether in the valleys of India or in the forests of Germany, by the common names of the Deity, all originally expressive of natural powers. Their worship is not, as has been so often said, a worship of nature. But if it had to be characterized by one word, I should venture to call it a worship of *God in Nature*, of God as appearing behind the gorgeous veil of Nature, rather than as hidden behind the veil of the sanctuary of the human heart. The gods of the Aryan pantheon assume an individuality so strongly marked and permanent, that with the Aryans, a transition to monotheism required a powerful struggle, and seldom took effect without iconoclastic revolutions or philosophical despair.

These three classes of religion are not to be mistaken, as little as the three classes of language, the Turanian, the Semitic, and the Aryan. They mark three events in the most ancient history of the world, events which have determined the whole fate of the human race, and of which we ourselves still feel the consequences in our language, in our thoughts, and in our religion.

But the chaos which these three heroes in language, thought, and religion, the Turanian, the Semitic, and the Aryan, left behind, was not altogether a chaos. The stream of language from which these three channels had separated rolled on; the sacred fire of religion from which these three altars had been lighted was not extinguished, though hidden in smoke and ashes. There was language and there was religion everywhere in the world, but it was natural, wild-growing language and religion; it had no history, it left no history, and it is therefore incapable of that peculiar scientific treatment which has been found applicable to a study of the languages and the religions of the Chinese, the Semitic, and the Aryan nations.

People wonder why the students of language have not succeeded in establishing more than three families of speech — or rather two, for the Turanian can hardly be called a family, in the strict sense of that word, until it has been fully proved that Chinese forms the centre of the two Turanian branches, the North Turanian on one side, and the South Turanian on the other; that Chinese[1] forms, in fact, the earliest settlement of that unsettled mass of speech, which, at a

[1] *Lecture on the Stratification of Language*, p. 4.

later stage, became more fixed and traditional: in the north, in *Tungusic, Mongolic, Tataric,* and *Finnic;* and in the south, in *Taic, Malaic, Bhotïya,* and *Tamulic.* Now the reason why scholars have discovered no more than these two or three great families of speech is very simple. There were no more, and we cannot make more. Families of languages are very peculiar formations; they are, and they must be, the exception, not the rule, in the growth of language. There was always the possibility, but there never was, as far as I can judge, any necessity of human speech leaving its primitive stage of wild growth and wild decay. If it had not been for what I consider a purely spontaneous act on the part of the ancestors of the Semitic, Aryan, and Turanian races, all languages might forever have remained ephemeral, answering the purposes of every generation that comes and goes, struggling on, now gaining, now losing, sometimes acquiring a certain permanence, but after a season breaking up again, and carried away like blocks of ice by the waters that rise underneath the surface. Our very idea of language would then have been something totally different from what it is now. For what are we doing? We first form our idea of what language ought to be from those exceptional languages which were arrested in their natural growth by social, religious, political, or at all events by extraneous influences, and we then turn round and wonder why all languages are not like these two or three exceptional channels of speech. We might as well wonder why all animals are not domesticated, or why, besides the garden anemone, there should be endless varieties of the same flower growing wild on the meadow and in the woods.

In the Turanian class, in which the original concentration was never so powerful as in the Aryan and Semitic families, we can still catch a glimpse of the natural growth of language, though confined within certain limits. The different settlements of this great floating mass of homogeneous speech do not show such definite marks of relationship as Hebrew and Arabic, Greek and Sanskrit, but only such sporadic coincidences and general structural similarities as can be explained by the admission of a primitive concentration, followed by a new period of independent growth. It would be willful blindness not to recognize the definite and characteristic features which pervade the North Turanian languages; it would be impossible to explain the coincidence between Hungarian, Lapponian, Esthonian, and Finnish, except on the supposition that there was a very early concentration of speech whence these dialects branched off. We see less clearly in the South Turanian group, though I confess my surprise even here has always been, not that there should be so few, but that there should be even these few relics, attesting the former community of these divergent streams of language. The point in which the South Turanian and North Turanian languages meet goes back as far as Chinese; for that Chinese is at the root of Mandshu and Mongolian as well as of Siamese and Tibetan becomes daily more apparent through the researches of Mr. Edkins. There is no hurry for pronouncing definitely on these questions; only we must not allow the progress of free inquiry to be barred by dogmatic skepticism; we must not look for evidence which from the nature of the case we cannot and ought not to find; and, before all things, we must not allow ourselves to be persuaded that for the discovery of truth, blinkers are more useful than spectacles.

If we turn away from the Asiatic continent, the original home of the Aryan, the Semitic, and the Turanian languages, we find that in Africa, too, a comparative study of dialects has clearly proved a concentration of African language, the results of which may be seen in the uniform *Bantu* dialects, spoken from the equator to the Keiskamma.[1] North of this body of Bantu or Kafir speech, we have an independent settlement of Semitic language in the Berber and the Galla dialects; south of it we have only the Hottentot and Bushman tongues, the latter hardly analyzed as yet, the former supposed to be related to languages spoken in Northern Africa, from which it became separated by the intrusion of the Kafir tribes. Some scholars have indeed imagined a relationship between the language of the Hottentots, the Nubian dialects, and the ancient Egyptian, a language which, whatever its real relationship may be, marks at all events another primeval settlement of speech and religion, outside the Asiatic continent. But while the spoken languages of the African continent enable us to see the general articulation of the primitive population of Africa, — for there is a continuity in language which nothing can destroy, — we know, and can know, but little of the growth and decay of African religion. In many places Mohammedanism and Christianity have swept away every recollection of the ancient gods; and even when attempts have been made by missionaries or travellers to describe the religious status of Zulus or Hottentots, they could only see the most recent forms of African faith, and those were changed almost invariably into grotesque caricatures. Of ancient African religion we have but one record, namely, in the monu-

[1] Bleek, *Comparative Grammar of the South African Languages*, p. 2.

ments of Egypt; but here, in spite of the abundance of materials, in spite of the ruins of temples, and numberless statues, and half-deciphered papyri, I must confess that we have not yet come very near to the beatings of the heart which once gave life to all this strange and mysterious grandeur.[1]

What applies to Africa applies to America. In the North we have the languages as witnesses of ancient migrations, but of ancient religion we have, again, hardly anything. In the South we know of two linguistic and political centres; and there, in Mexico and Peru, we meet with curious, though not always trustworthy, traditions of an ancient and well-established system of religious faith and worship.

The Science of Religion has this advantage over the Science of Language, if advantage it may be called, that in several cases where the latter has materials sufficient to raise problems of the highest importance, but not sufficient for their satisfactory solution, the former has no materials at all. The ancient temples are destroyed, the names of the ancient deities are clean forgotten in many parts of the world where dialects, however changed, still keep up the tradition of the most distant ages. But even if it were otherwise, the students of religion would, I think, do well to follow the example of the students of language, and to serve their first apprenticeship in a comparative study of the Aryan and Semitic religions. If it can only be proved that the religions of the Aryan nations are united by the same bonds of a real relationship which have enabled us to treat their languages as so many varieties of the same type; and so also of the Semitic;

[1] De Vogue, *Journal Asiatique*, 1867, p. 136. De Rouge, "Sur la Religion des Anciens Egyptiens," in *Annales de Philosophie Chretienne*, Nov. 1869.

the field thus opened is vast enough, and its careful clearing and cultivation will occupy several generations of scholars. And this original relationship, I believe, can be proved. Names of the principal deities, words also expressive of the most essential elements of religion, such as *prayer, sacrifice, altar, spirit, law,* and *faith,* have been preserved among the Aryan and among the Semitic nations, and these relics admit of one explanation only. After that, a comparative study of the Turanian religions may be approached with better hope of success; for that there was not only a primitive Aryan and a primitive Semitic religion, but likewise a primitive Turanian religion, before each of these primeval races was broken up and became separated in language, worship, and national sentiment, admits, I believe, of little doubt.

Let us begin with our own ancestors, the Aryans. In a lecture which I delivered in this place some years ago, I drew a sketch of what the life of the Aryans must have been before their first separation, that is, before the time when Sanskrit was spoken in India, or Greek in Asia Minor and Europe. The outline of that sketch and the colors with which it was filled were simply taken from language. We argued that it would be possible, if we took all the words which exist in the same form in French, Italian, and Spanish, to show what words, and therefore what things, must have been known to the people who did not as yet speak French, Italian, and Spanish, but who spoke that language which preceded these Romance dialects. We happen to know that language; it was Latin; but if we did not know a word of Latin or a single chapter of Roman history, we should still be able, by using the evidence of the words which are common to all the

Romance languages, to draw some kind of picture of what the principal thoughts and occupations of those people must have been who lived in Italy a thousand years at least before the time of Charlemagne. We could easily prove that those people must have had *kings* and *laws, temples* and *palaces, ships* and *carriages, high roads* and *bridges,* and nearly all the ingredients of a highly civilized life. We could prove this, as I said, by simply taking the names of all these things as they occur in French, Spanish, and Italian, and by showing that as Spanish did not borrow them from French, or Italian from Spanish, they must have existed in that previous stratum of language from which these three modern Romance dialects took their origin.

Exactly the same kind of argument enabled us to put together a kind of mosaic picture of the earliest civilization of the Aryan people before the time of their separation. As we find in Greek, Latin, and Sanskrit, also in Slavonic, Celtic, and Teutonic, the same word for " *house,*" we are fully justified in concluding that before any of these languages had assumed a separate existence, a thousand years at least before Agamemnon and before Manu, the ancestors of the Aryan race were no longer dwellers in tents, but builders of permanent houses.[1] As we find the name for town the same in Sanskrit and Greek,[2] we can conclude with equal certainty that towns were known to the Aryans before Greek and before Sanskrit was spoken. As we find the name for king the same in Sanskrit, Latin, Teutonic, and Celtic,[3] we know again that kingly gov-

---

[1] Sk. dama, δόμος, domus, Goth. tim rjan, "to build," Sl. dom. Sk. vesa, οἶκος, vicus, Goth. veih-s.

[2] Sk. pur, puri, or puri; Gr. πόλις; Sk. vastu, "house"; Gr. ἄστυ.

[3] Sk. Raŷ, ragan, rex; Goth. reiks; Ir ━━━

ernment was established and recognized by the Aryans at the same prehistoric period. I must not allow myself to be tempted to draw the whole of that picture of primeval civilization over again.[1] I only wish to call back to your recollection the fact that in exploring together the ancient archives of language, we found that the highest god had received the same name in the ancient mythology of India, Greece, Italy, and Germany, and had retained that name whether worshipped on the Himalayan mountains, or among the oaks of Dodona, on the Capitol, or in the forests of Germany. I pointed out that his name was *Dyaus* in Sanskrit, *Zeus* in Greek, *Jovis* in Latin, *Tiu* in German; but I hardly dwelt with sufficient strength on the startling nature of this discovery. These names are not mere names; they are historical facts, aye, facts more immediate, more trustworthy, than many facts of mediæval history. These words are not mere words, but they bring before us, with all the vividness of an event which we witnessed ourselves but yesterday, the ancestors of the whole Aryan race, thousands of years it may be before Homer and the Veda, worshipping an unseen Being, under the selfsame name, the best, the most exalted name, they could find in their vocabulary, — under the name of Light and Sky. And let us not turn away, and say that this was after all but nature-worship and idolatry. No, it was not meant for that, though it may have been degraded into that in later times; *Dyaus* did not mean the blue sky, nor was it simply the sky personified — it was meant for something else. We have in the Veda the invocation *Dyaus pitar*, the Greek Ζεῦ πάτερ, the Latin *Jupiter*; and that means in all the three languages what it

---

[1] See *Chips from a German Workshop*, vol. ii. p. 22, seq.

meant before these three languages were torn asunder — it means Heaven-Father! These two words are not were words; they are to my mind the oldest poem, the oldest prayer of mankind, or at least of that pure branch of it to which we belong, — and I am as firmly convinced that this prayer was uttered, that this name was given to the unknown God before Sanskrit was Sanskrit and Greek was Greek, as, when I see the Lord's Prayer in the languages of Polynesia and Melanesia, I feel certain that it was first uttered in the language of Jerusalem. We little thought when we heard for the first time the name of Jupiter, degraded it may be by Homer or Ovid into a scolding husband or a faithless lover, what sacred records lay enshrined in this unholy name. We shall have to learn the same lesson again and again in the Science of Religion, namely, that the place whereon we stand is holy ground. Thousands of years have passed since the Aryan nations separated to travel to the North and the South, the West and the East; they have each formed their languages, they have each founded empires and philosophies, they have each built temples and razed them to the ground; they have all grown older, and it may be wiser and better; but when they search for a name for what is most exalted and yet most dear to every one of us, when they wish to express both awe and love, the infinite and the finite, they can but do what their old fathers did when gazing up to the eternal sky, and feeling the presence of a Being as far as far, and as near as near can be; they can but combine the selfsame words, and utter once more the primeval Aryan prayer, Heaven-Father, in that form which will endure forever, "Our Father which art in heaven."

Let us now turn to the early religion of the Semitic

nations. The Semitic languages, it is well known, are even more closely connected together than the Aryan languages, so much that a comparative grammar of the Semitic languages seems to have but few of the attractions possessed by a comparative study of Sanskrit, Greek, and Latin. Semitic scholars complain that there is no work worth doing in comparing the grammars of Hebrew, Syriac, Arabic, and Ethiopic, for they have only to be placed side by side [1] in order to show their close relationship. I do not think this is quite the case, and I still hope that M. Renan will carry out his original design, and, by including not only the literary branches of the Semitic family, but also the ancient dialects of Phenicia, Arabia, Babylon, and Nineveh, produce a comparative grammar of the Semitic languages that may hold its place by the side of Bopp's great work on the "Comparative Grammar of the Aryan Languages."

But what is still more surprising to me is that no Semitic scholar should have followed the example of the Aryan scholars, and collected from the different Semitic dialects those common words which must have existed before Hebrew was Hebrew, before Syriac was Syriac, and before Arabic was Arabic, and from which some kind of idea might be formed as to what were the principal thoughts and occupations of the Semitic race in its earliest undivided state. The materials seem much larger and much more easily accessible.[3] The principal degrees of relationship, for instance, have common names among the Semitic as among the Aryan nations, and if it was important to show that the Aryans had named and recognized not only the natural

---

[1] See Bunsen's *Christianity and Mankind*, vol. iii. p. 246, *seq.*
[2] Ibid. iii. 246; iv. 345.

members of a family, such as father and mother, son and daughter, brother and sister, but also the more distant members, the father and mother-in-law, the son and daughter-in-law, the brother and sister-in-law, would it not be of equal interest to show that the Semitic nations had reached the same degree of civilization long before the time of the laws of Moses?

Confining ourselves to the more immediate object of our researches, we see without difficulty that the Semitic languages, like the Aryan languages, possess a number of names of the Deity in common, which must have existed before the *Southern* or *Arabic*, the *Northern* or *Aramaic*, the *Middle* or *Hebraic* branches became permanently separated, and which, therefore, allow us an insight into the religious conceptions of the once united Semitic race long before Jehovah was worshipped by Abraham, or Baal was invoked in Phenicia, or El in Babylon.

It is true, as I pointed out before, that the meaning of many of these names is more general than the original meaning of the names of the Aryan gods. Many of them signify *Powerful*, *Venerable*, *Exalted*, *King*, *Lord*, and they might seem, therefore, like honorific titles, to have been given independently by the different branches of the Semitic family to the gods whom they worshipped each in their own sanctuaries. But if we consider how many words there were in the Semitic languages to express greatness, strength, or lordship, the fact that the same appellatives occur as the proper names of the deity in Syria, in Carthage, in Babylon, and in Palestine, admits of one historical explanation only. There must have been a time for the Semitic as well as for the Aryan races, when they fixed the names of their deities, and that time must have preceded the formation of their separate languages and separate religions.

One of the oldest names of the deity among the ancestors of the Semitic nations was *El*. It meant Strong. It occurs in the Babylonian inscriptions as Ilu, God,[1] and in the very name of Bab-il, the gate or temple of Il. In Hebrew it occurs both in its general sense of strong or hero, and as a name of God; and we find it applied, not to the true God only, but also to the gods of the Gentiles or to false gods. We have it in *Beth-el*, the house of God, and in many other names. If used with the article, as *ha-El*, the Strong One, or the God, it always is meant in the Old Testament for Jehovah, the true God.

The same El was worshipped at Byblus by the Phenicians, and he was called there the son of Heaven and Earth.[2] His father was the son of *Eliun*, the most high God, who had been killed by wild animals. The son of Eliun, who succeeded him, was dethroned, and at last slain by his own son *El*, whom Philo identifies with the Greek Kronos, and represents as the presiding deity of the planet Saturn.[3] In the Himyaritic inscriptions, too, the name of El has been discovered.[4]

With the name of *El*, Philo connected the name of *Elohim*, the plural of *Eloah*. In the battle between *El* and his father, the allies of *El*, he says, were called

---

[1] Schrader, in *Zeitschrift der Deutschen Morgenländischen Gesellschaft*, vol. xxiii. p. 350.

[2] Bunsen, *Egypt*, iv. 187. *Fragmenta Hist. Græc.*, vol. iii. p. 567.

[3] *Fragmenta Hist. Græc.*, iii. pp. 567–571. That El is the presiding deity of the planet Saturn according to the Chaldæans is also confirmed by Diodorus Siculus, ii. 30–3.

[4] Osiander, *Zeitschrift der Deutschen Morgenländischen Gesellschaft*, x. 61.

Eloeim, as those who were with *Kronos* were called *Kronioi*.[1] This is, no doubt, a very tempting etymology of *Eloah;* but as the best Semitic scholars, and particularly Professor Fleischer, have declared against it, we shall have, however reluctantly, to surrender it. Eloah is the same word as the Arabic *ilah*, God.

In the singular, *Eloah* is used in the Bible synonymously with *El;* in the plural it may mean gods in general or false gods, but it becomes in the Old Testament the recognized name of the true God, plural in form, but singular in meaning. In Arabic, *ilah*, without the article, means a God in general; with the article, Al-Ilah, or Allah, becomes the name of the God of Mohammed, as it was the name of the God of Abraham and of Moses.

The origin of *Eloah* or *Ilah* has been frequently discussed by European as well as by native scholars. The Kamus says that there were twenty, Mohammed El Fasi that there were thirty, opinions about it. Professor Fleischer,[2] whose judgment in such matters we may trust implicitly, traces *El*, the strong one, back to a root *al* (with middle vav, aval), to be thick and dense, to be fleshy and strong. But he takes *Eloah* or *Ilah* for an abstract noun, in the sense of fear, derived from a totally different root, namely, *alah*, to be agitated, confounded, perplexed. From meaning fear, *Eloah* came to mean the object of fear or reverence,

---

[1] *Frag. Hist. Græc.*, iii. 568, 18. οἱ δὲ σύμμαχοι Ἤλου τοῦ Κρόνου Ἐλοείμ ἐπεκλήθησαν, ὡς ἂν Κρόνιοι οὗτοι ἦσαν οἱ λεγόμενοι ἐπὶ Κρονου.

[2] See a note by Professor Fleischer in Delitzsch, *Commentar uber die Genesis*, third ed. 1860, p. 64; also *Zeitschrift der Deutschen Morgenlandischen Gesellschaft*, vol. x. p. 60; and *Sitzungsberichte der konigl. Sachsischen Gesellschaft der Wissenschaften, Philosoph. Hist. Classe*, vol. xviii. (1866), pp. 290-292. Dr. W. Wright adopts Prof. Fleischer's derivation; likewise Prof. Kuenen, in his work, *De Godsdienst van Israel*, p. 45.

and thus became a name of God. In the same way we find *pachad*, which means fear, used in the sense of God; Gen. xxxi. 42: "Except the God of my father, the God of Abraham, and the fear of Isaac had been with me." And again, v. 53: "And Jacob sware by the fear of his father Isaac." In Aramaic, *dachla*, fear, is the recognized name for God or for an idol.

The same ancient name appears also in its feminine form as *Allat*.[1] Her famous temple at *T*aif, in Arabia, was second only in importance to the sanctuary of Mekkah, and was destroyed at the command of Mohammed. The worship of *Allat*, however, was not confined to this one place; and there can be no doubt that the Arabian goddess *Alibat*, mentioned by Herodotus,[2] is the same as the *Allat* of the Koran.

Another famous name of the deity, traces of which can be found among most of the Semitic nations, is *Baal* or *Bel*. The Assyrians and Babylonians,[3] the Phenicians[4] and Carthaginians, the Moabites and Philistines, and, we must add, the Jews also, all knew of *Bel* or *Baal* as a great, or even as the supreme God. Baal can hardly be considered as a strange and foreign god in the eyes of the Jewish people, who in spite of the protests of the Hebrew prophets, worshipped him so constantly in the groves of Jerusalem. He was felt by them almost as a home deity, or, at all events, as a

---

[1] Osiander, *Zeitschrift der Deutschen Morgenlandischen Gesellschaft*, vii. pp. 479–482. Allat, goddess, is contracted from Al-Ilahat.

[2] Herod. iii. 8. Ὀνομάζομαι (οἱ Ἀράβιοι) τὸν μὲν Διόνυσον Ὀροτάλ, τὴν δὲ Οὐρανίμν Ἀλιλάτ. In Herod. i. 131, 138, this name is corrupted to Ἀλιττα. See Osiander, *Zeitschrift der Deutschen Morgenlandischen Gesellschaft*, ii. 482, 483.

[3] *Fragmenta Hist. Græc.*, ii. 498, 2.

[4] Ibid. iii. 568, 21.

Semitic deity, and among the gods whom the fathers served on the other side of the flood, Bel held most likely a very prominent place. Though originally *one*,[1] Baal became divided into many divine personalities through the influence of local worship. We hear of a Baal-tsur, Baal-tsidon, Baal-tars, originally the Baal of Tyre, of Sidon, and Tarsus. On two candelabra found on the island of Malta we read the Phenician dedication to " Melkarth, the Baal of Tyre." At Shechem Baal was worshipped as *Baal-barith*,[2] supposed to mean the god of treaties; at Ekron the Philistines worshipped him as *Baal-zebub*,[3] the lord of flies ; while the Moabites, and the Jews too, knew him also by the name of *Baal-peor*.[4] On Phenician coins Baal is called Baal Shamayîm, the Baal of heaven, which is the *Beelsamen* of Philo, identified by him with the sun.[5] " When the heat became oppressive, the ancient races of Phenicia," he says, " lifted their hand heavenward to the sun. For him they considered the only God, the lord of heaven, calling him Beelsamên,[6] which with the Phenicians is lord of heaven, and with the Greeks Zeus." We likewise hear of *Baalim*, or many Baals or gods. And in the same way as by the side of the male *Ilah* or *Allah* we found a female *Allat*, we also find by the side of the male Baal, a female deity *Baalt*, the Baaltis of the Phenicians. It may be that the original conception of female deities differs among

[1] M. de Vogue, *Journal Asiatique*, 1867, p. 135.
[2] Judges viii. 33 ; ix. 4.   [3] 2 Kings i. 2, 3, 16.   [4] Numbers xxv.
[5] *Fragmenta Hist. Græc.* iii. 565, 5. It is impossible to change ἥλιον into ἥλον, because El or Kronos is mentioned afterwards.
[6] Is this the same as Barsamus, mentioned by Moses of Chorene (*Hist. Arm.* i. 13) as a deified hero worshipped by the Syrians ? Or is Barsamus the Son of Heaven ? See Rawlinson, *Ancient Monarchies*, vol. i. p. 116.

Semitic and Aryan nations, and that these feminine forms of *Allah* and *Baal* were at first intended only to express the energy or activity, or the collective powers of the deity, not a separate being, least of all a wife. This opinion [1] is certainly confirmed when we see that in a Carthaginian inscription the goddess *Tanit* is called *the face of Baal*, and that in the inscription of Eshmunazar, the Sidonian Astarte is called the *name of Baal*. In course of time, however, this abstract idea was supplanted by that of a female power, and even a wife, and as such we find *Baaltis* worshipped by Phenicians,[2] Babylonians, and Assyrians;[3] for the name of Mylitta in Herodotus [4] is, according to Dr. Oppert, a mere corruption of Baaltis.

Another famous female goddess is *Ashtoreth*, a name which presupposes a masculine deity, *Ashtar*. Traces of this god have been discovered in the *Ishtar* of the Babylonian inscriptions, and more recently in the Ashtar of the Moabite stone. In this case, however, the female deity became predominant, and was worshipped, not only by Carthaginians, Phenicians, and Philistines, but likewise by the Jews,[5] when they forsook the Lord, and served Baal and Ashtaroth.[6] The Syrians called her Astarte, and by that ominous name she became known to Greeks and Romans. When Jeremiah speaks of the Queen of Heaven,[7] this can only be meant for Astarte, or Baaltis. Even in Southern Arabia there are traces of the worship of this ancient goddess. For in Sanâ, the ancient capital of the Himyaritic kingdom, there was a magnificent palace and temple dedicated to Venus (Bait Ghumdân), and the

---

[1] De Vogue, l. c. p. 138. [2] *Fragmenta Hist. Græc.* iii. 569, 25.
[3] Ibid. iv. 283, 9. [4] Herod. i. 131, 199. [5] 1 Kings xi. 5.
[6] Judges iii. 12. [7] Jer. vii. 18.

name of *Athtar* has been read in the Himyaritic inscriptions: nay, it is preceded in one place by the verb in the masculine gender.[1]

Another word, meaning originally king, which must have been fixed upon as a name of the Deity in prehistoric times, is the Hebrew *Melech*. We find it in *Moloch*, who was worshipped, not only at Carthage, in the Islands of Crete and Rhodes, but likewise in the valley of Hinnom. We find the same word in *Milcom*, the god of the Ammonites, who had a sanctuary in Mount Olivet; and the gods *Adrammelech* and *Anammelech*, to whom the Sepharvites burnt their children in the fire,[2] seem again but local varieties of the same Semitic idol.

Adonai, which in Hebrew means my lord, and in the Old Testament is used exclusively of Jehovah, appears in Phenicia as the name of the Supreme Deity, and after undergoing manifold mythological transformations, the same name has become familiar to us through the Greek tales about the beautiful youth Adonis, loved by Aphrodite, and killed by the wild boar of Ares.

Elyon, which in Hebrew means the Highest, is used in the Old Testament as a predicate of God. It occurs also by itself as a name of Jehovah. Melchizedek is called emphatically the priest of *El elyon*, the priest of the most high God.

But this name, again, is not restricted to Hebrew. It occurs in the Phenician cosmogony as *Eliun*, the highest God, the Father of Heaven, who was the

---

[1] Osiander, l. c. p. 472; Gildemeister, *Zeitschrift der D. M. G.* xxiv. pp. 180, 181; Lenormant, *Comptesrendus des seances de l'Acad. des incriptions et belles-lettres de l'annee* 1867; Levy, *Zeitschrift der D M. G.* xxiv. p. 189.

[2] 2 Kings xvii. 31.

father of *El.* Dr. Oppert has identified this Eliun with the *Illinus* mentioned by Damascius.

Another word used in the Bible, sometimes in combination with El, and more frequently alone, as a name of the supreme deity, is *Shaddai,* the Powerful. It comes from a kindred root to that which has yielded the substantive *Shed,* meaning demon in the language of the Talmud, and the plur. *Shedim,* a name for false gods or idols in the Old Testament. This name occurs as *Set* or *Sed* in the hieroglyphic inscriptions.[1] It is there the name of a god introduced by the shepherds, and one of his surnames is given as *Baal.* The same deity *Shaddai,* the Powerful, has, by a clever conjecture, been discovered as one of the deities worshipped by the ancient Phenicians.[2]

While these names of the Deity and some others are shared in common by all, or by the most important members of the Semitic family, and must therefore have existed previous to the first Semitic separation, there are others peculiar to each branch.

Thus the name of Jehovah, or *Jahveh,*[3] as it seems originally to have been pronounced, seems to me to be a divine name peculiar to the Jews. It is true that in a well-known passage of Lydus, IAO[4] is said to have been the name of God among the Chaldæans. But granting that IAO was the same word as Jahveh

---

[1] De Vogué, l. c. p. 160.
[2] Bunsen, *Egypt,* iv. 221. De Vogué, *Mélanges d'Archeologie,* p. 77.
[3] Theodoret. *Quest. xv. ad Exodum* (420 A. D.): καλοῦσι δὲ αὐτο Σαμαρεῖται ΙΑΒΕ, Ἰουδαῖοι δὲ ΙΑΩ. Diod. Sic. i. 94 (59 B. C.): παρὰ δὲ τοῖς Ἰουδαίοις Μωυσῆν τὸν Ἰαὼ ἐπικαλούμενον θεόν, κ. τ. λ.
[4] Lydus, *De Mensibus,* iv. 38, 14: Οἱ Χαλδαῖοι τὸν θεὸν ΙΑΩ λέγουσι, ἀντὶ, τοῦ φῶς νοητον· τῇ Φοινίκων γλώσσῃ καὶ ΣΑΒΑΩΘ δὲ πολλαχοῦ λέγεται, οἷον ὁ ὑπερ τοὺς επτς πολους, τουτέστιν ὁ δημιουργος. Bunsen, *Egypt,* iv. 193; Renan, *Sanchoniathon,* p. 44, *note.* And see Diodorus Siculus, i. 94, 2.

or Jehovah or Jah (as in Hallelu-jah), may not Lydus by the Chaldæans have simply meant the Jews? If, as Sir Henry Rawlinson maintains, the name of Jehovah occurred in the Babylonian inscriptions, the case would be different; we should then have to admit that this name, too, was fixed before the Semitic family was broken up; but until this is fully proved, we shall be justified in claiming *Jehovah* as a name of the Deity peculiar to Hebrew, or, at all events, as fixed by the Hebrew prophets in the sense of the one true God, opposed to all other gods of the Semitic race.[1]

But whether we include or exclude the name of Jehovah, we have, I think, sufficient witnesses to establish what we wished to establish, namely, that there was a period during which the ancestors of the Semitic family had not yet been divided, whether in language or in religion. That period transcends the recollection of every one of the Semitic races in the same way as neither Hindus, Greeks, nor Romans have any recollection of the time when they spoke a common language, and worshipped their Father in heaven by a name that was as yet neither Sanskrit, nor Greek, nor Latin. But I do not hesitate to call this pre-historic period historical in the best sense of the word. It was a real period, because, unless it was real, all the realities of the Semitic languages and the Semitic religions, such as we find them after their separation, would be unintelligible. Hebrew, Syriac, and Arabic point to a common source as much as Sanskrit, Greek, and Latin; and unless we can bring ourselves to doubt

---

[1] Lobeck, *Aglaophamus*, p. 461. Sir H. Rawlinson has kindly informed me that he doubts whether Yahu, which occurs in the sense of God in the Assyrian inscription, belonged properly to the Assyrian language. He thinks that it may have been borrowed from Syria, and adopted with the language, as were so many other foreign terms.

that the Hindus, the Greeks, the Romans, and the Teutons derived the worship of their principal deity from their common Aryan sanctuary, we shall not be able to deny that there was likewise a primitive religion of the whole Semitic race, and that *El*, the Strong One in Heaven, was invoked by the ancestors of all the Semitic races, before there were Babylonians in Babylon, Phenicians in Sidon and Tyrus, before there were Jews in Mesopotamia or Jerusalem. The evidence of the Semitic is the same as that of the Aryan languages; the conclusion cannot be different.

We now come to the third nucleus of language, and as I hope to show, of religion also, — to that which forms the foundation of the Turanian world. The subject is extremely difficult, and I confess I doubt whether I shall succeed in engaging your sympathy in favor of the religious opinions of people so strange, so far removed from us, as the Chinese, the Mongolians, the Samoyedes, the Finns, and Lapps. We naturally take an interest in the ancient history of the Aryan and Semitic nations, for, after all, we are ourselves Aryan in language, and Semitic, at least to a certain extent, in religion. But what have we in common with the Turanians, with Chinese and Samoyedes? Very little, it may seem; and yet it is not very little, for it is our common humanity. It is not the yellow skin and the high cheek-bones that make the man. Nay, if we look but steadily into those black Chinese eyes, we shall find that there, too, there is a soul that responds to a soul, and that the God whom they *mean* is the same God whom we *mean*, however helpless their utterance, however imperfect their worship.

If we take the religion of China as the earliest representative of Turanian worship, the question is, whether we can find any names of the Deity in Chinese which appear again in the religions and mythologies of other Turanian tribes, such as the Mandshus, the Mongolians, the Tatars, or Finns. I confess that, considering the changing and shifting character of the Turanian languages, considering also the long interval of time that must have passed between the first linguistic and religious settlement in China, and the later gradual and imperfect consolidation of the other Turanian races, I was not very sanguine in my expectation that any such names as *Dyaus Pitar* among the Aryans, or *El* and *Baal* among the Shemites, could have survived in the religious traditions of the vast Turanian world. However, there is no reason why we should not look for such names in Chinese, Mongolian, and Turkish; still less, why we should pass them by with indifference or incredulity, because, from the very nature of the case, their coincidence is not so striking and convincing as that of the Semitic and Aryan names of the Deity. There are in researches of this kind different degrees of certainty, and I am the very last person to slur them over, and to represent all our results as equally certain. But if we want to arrive at *terra firma*, we must not mind a plunge now and then; and if we wish to mount a ladder, we must not be afraid of taking the first step. The coincidences between the religious phraseology of Chinese and other Turanian languages are certainly not like the coincidences between Greek and Sanskrit, or between Hebrew and Phenician; but they are such that they ought not to be neglected by the pioneers of a new science.

You remember that the popular worship of ancient China was a worship of single spirits, of powers, or, we might almost say, of names; the names of the most prominent powers of nature which are supposed to exercise an influence for good or evil on the life of man. We find a belief in spirits of the sky, the sun, the moon, the stars, the earth, the mountains, the rivers; to say nothing as yet of the spirits of the departed. In China, where there always has been a strong tendency towards order and regularity, some kind of system has been superinduced by the recognition of two powers, one active, the other passive, one male, the other female, which comprehend everything, and which, in the mind of the more enlightened, tower high above the great crowd of minor spirits. These two powers are within and beneath and behind everything that is double in nature, and they have frequently been identified with heaven and earth. We can clearly see, however, that the spirit of heaven occupied from the beginning a much higher position than the spirit of the earth. It is in the historical books only, in the Shu King,[1] that we are told that heaven and earth together are the father and mother of all things. In the ancient poetry *Heaven* alone is both father and mother.[2] This spirit of heaven is known in Chinese by the name of *Tien*, and wherever in other religions we should expect the name of the supreme deity, whether Jupiter or Allah, we find in Chinese the name of *Tien* or sky. This *Tien*, according to the Imperial Dictionary of Kanghee, means the

---

[1] In the Shu-king (3, 11) Tien is called Shang-tien, or High Heaven, which is synonymous with Shang-te, High Spirit, another very common name of the supreme deity. The Confucians never made any image of Shang-te, but the Taosse represented their (Yah-hwang) Shang-te under the human form. Medhurst, *Inquiry*, p. 46.

[2] Chalmers, *Origin of the Chinese*, p 14; Medhurst, l. c. p. 124; contrast between Shins and Shangti.

Great One, he that dwells on high and regulates all below. We see in fact that *Tien*, originally the name of the sky, has passed in Chinese through nearly all the phases, from the lowest to the highest, through which the Aryan name for sky, *dyaus*, passed in the poetry, the religion, mythology, and philosophy of India and Greece. The sign of tien in Chinese is compounded of two signs: *ta*, which means *great*, and *yih*, which means *one*. The sky, therefore, was conceived as the One, the Peerless, and as the Great, the High, the Exalted. I remember reading in a Chinese book, "As there is but one sky, how can there be many gods?" In fact, their belief in *Tien*, the spirit of heaven, moulded the whole of the religious phraseology of the Chinese. "The glorious heaven," we read, "is called bright, it accompanies you wherever you go; the glorious heaven is called luminous, it goes wherever you roam." *Tien* is called the ancestor of all things; the highest that is above. He is called the great framer, who makes things as a potter frames an earthen vessel. The Chinese also speak of the decrees and the will of Heaven, of the steps of Heaven or Providence. The sages who teach the people are sent by Heaven, and Confucius himself is said to have been used by Heaven as the "alarum" of the world. The same Confucius, when on the brink of despondency, because no one would believe in him, knows of one comfort only; that comfort is, "Heaven knows me." It is clear from many passages that with Confucius Tien or the Spirit of Heaven was the supreme deity, and that he looked upon the other gods of the people, the spirits of the air, the mountains, and the rivers, the spirits also of the departed, very much with the same feeling with which Sokrates regarded the mythological deities of Greece. Thus when asked on one occasion how the spirits should be served, he replied, " If we are not able to serve men, how can we serve the spirits?" And at another time he said in his short and significant manner, " Respect the Gods, and keep them at a distance." [1]

We have now to see whether we can find any traces of this belief in a supreme spirit of heaven among the other branches of the Turanian class, the Mandshus, Mongolians, Tatars, Finns, or Lapps. As there are many names for sky in the Turanian dialects, it would not be absolutely necessary that we should find the same name which we found in Chinese; yet, if traces of that name could be found among Mongolians and Tatars, our argument would, no doubt, gain far greater strength. It is the same in all researches of comparative mythology. If we find the same conceptions, the same myths and legends, in India, Greece, Italy, and Germany, there is, no doubt, some presumption in favor of their common origin, but no more. But if we meet with gods and heroes, having the same name in the mythology of the Veda, and the mythology of Greece and Rome and Germany, then we stand on firmer ground. We have then to deal with real facts that cannot be disputed, and all that remains is to explain them. In Turanian mythology, however, such facts are not easily brought together. With the exception of China, we know very little of the ancient history of the Turanian races, and what we know of their present state comes frequently from prejudiced observers. Besides, their old heathendom is fast disappearing before the advance of Buddhism, Mohammedanism, and Christianity. Yet if we take the accounts of the most

[1] Medhurst, *Reply to Dr. Boone*, p. 32.

trustworthy travellers in Central and Northern Asia, and more particularly the careful observations of Castrén, we cannot but recognize some most striking coincidences in the scattered notices of the religion of the Tungusic, Mongolic, Tataric, and Finnic tribes. Everywhere we find a worship of the spirits of nature, of the spirits of the departed, though, behind and above it there rises the belief in some higher power, known by different names, sometimes called the Father, the Old One, who is the Maker and Protector of the world, and who always resides in heaven. Chinese historians are the only writers who give us an account of the earlier history of some of these Turanian tribes, particularly of the Huns, whom they call *Hiongnu*, and of the Turks, whom they call *Tukiu*. They relate[1] that the Huns worshipped the sun, the moon, the spirits of the sky and the earth, and the spirits of the departed, and that their priests, the Shamans, possessed a power over the clouds, being able to bring down snow, hail, rain, and wind.[2]

Menander, a Byzantine historian, relates of the Turks that in his time they worshipped the fire, the water, and the earth, but that at the same time they believed in a God, the maker of the world, and offered to him sacrifices of camels, oxen, and sheep.

Still later we get some information from mediæval travellers, such as Plano Carpini, and Marco Polo, who say that the Mongol tribes paid great reverence to the sun, the fire, and the water, but that they believed also in a great and powerful God, whom they called *Natagai* (Natigay) or *Itoga*.

In modern times we have chiefly to depend on Cas-

---

[1] Castrén, *Vorlesungen ueber Finnische Mythologie*, p. 2.
[2] Ibid. l. c. p. 36.

trén, who had eyes to see and ears to hear what few other travellers would have seen or heard, or understood. Speaking of the Tungusic tribes, he says, "They worship the sun, the moon, the stars, the earth, fire, the spirits of forests, rivers, and certain sacred localities; they worship even images and fetiches, but with all this they retain a faith in a supreme being which they call *Buga*."[1] "The Samoyedes," he says, "worship idols and various natural objects; but they always profess a belief in a higher divine power which they call *Num*."

This deity which is called *Num*, is also called *Juma* by the Samoyedes,[2] and is in fact the same deity which in the grand mythology of Finland is known under the name of *Jumala*. The mythology of Finland has been more carefully preserved than the mythologies of all the other Altaic races, and in their ancient epic poems, which have been kept up by oral tradition for centuries, and have been written down but very lately, we have magnificent descriptions of Jumala, the deity of the sky. *Jumala* meant originally the sky. It is derived, as Castrén has shown,[3] from *Juma*, thunder, and *la*, the place, meaning therefore, the place of thunder, or the sky. It is used first of all for sky, secondly for god of the sky, and thirdly for gods in general. The very same word, only modified according to the phonetic rules of each language, occurs among the Lapps,[4] the Esthonians, the Syrjanes, the Tcheremissians, and the Votyakes.[5] We can watch the growth and the changes of this heavenly deity as we catch a glimpse here and there of the religious thoughts of these Altaic tribes. An old Samoyede woman who was asked by

---

[1] Is this the Russian "bog," god? [2] Castrén, l. c. p. 18.
[3] Page 24.   [4] Page 11.   [5] Page 24.

Castrén[1] whether she ever said her prayers, replied, "Every morning I step out of my tent and bow before the sun, and say: 'When thou risest, I, too, rise from my bed.' And every evening I say: 'When thou sinkest down, I too, sink down to rest.'" That was her prayer, perhaps the whole of her religious service, — a poor prayer it may seem to us, but not to her; for it made that old woman look twice at least every day away from earth and up to heaven; it implied that her life was bound up with a larger and higher life; it encircled the daily routine of her earthly existence with something of a divine halo. She herself was evidently proud of it, for she added, with a touch of self-righteousness, "There *are* wild people who never say their morning and evening prayers."

As in this case the deity of the sky is represented, as it were, by the sun, we see Jumala, under different circumstances, conceived as the deity of the sea. When walking one evening with a Samoyede sailor along the coast of the Polar Sea, Castrén asked him, "Tell me, where is Num?" (*i. e.* Jumala.) Without a moment's hesitation the old sailor pointed to the dark, distant sea, and said, "*He is there.*"

Again, in the epic poem Kalevála, when the hostess of Pohjola is in labor, she calls on Jumala, and says, "Come now into the bath, Jumala, into the warmth, O lord of the air!"[2]

At another time Jumala is the god of the air, and is invoked in the following lines:[3] —

> "Harness now thyself, Jumala,
> Ruler of the air, thy horses!
> Bring them forth, thy rapid racers,
> Drive the sledge with glittering colors,

[1] Page 16.   [2] Page 19.   [3] Page 21.

> Passing through our bones, our ankles,
> Through our flesh that shakes and trembles,
> Through our veins which seem all broken.
> Knit the flesh and bones together,
> Fasten vein to vein more firmly.
> Let our joints be filled with silver,
> Let our veins with gold be running!"

In all these cases the deity invoked is the same, it is the deity of the sky, Jumala; but so indefinite is his character, that we can hardly say whether he is the god of the sky, or the sun, or the sea, or the air, or whether he is a supreme deity reflected in all these aspects of nature.

However, you will naturally ask, where is there any similarity between the name of that deity and the Chinese deity of the sky, *Tien?* The common worship of *Jumala* may prove some kind of religious concentration among the different Altaic nations in the North of Asia, but it does not prove any pre-historic community of worship between those nations and the ancient inhabitants of China. It is true that the Chinese *Tien* with its three meanings of sky, god of the sky, and god in general, is the exact counterpart of the North Turanian Jumala; but still we want more; we want traces of the same name of the deity in China, in Mongolia, and Tatary, just as we found the name of Jupiter in India and Italy, and the name of El in Babylon and Palestine.

Well, let us remember that Chinese is a monosyllabic language, and that the later Turanian dialects have entered into the agglutinative stage, that is to say, that they use derivative suffixes, and we shall then without much difficulty discover traces of the Chinese word *Tien*, with all its meanings, among some at least of the most important of the Turanian races. In the Mongolian language we find *Teng-ri*,[1] and this means, first, sky; then, god of the sky; then, god in general; and lastly, spirit or demon, whether good or bad.

I think you will see the important bearing of this discovery, for it clinches the argument as nothing else could have clinched it. Unless we had found the same name of the supreme deity in the hymns of the Veda and in the prayer of the priestesses at Dodona, we could not have forced the conviction that it was originally one and the same conception of divine personality, that had been worshipped long before the Hindus had entered India, or the dove had alighted on the oak of Dodona. The same applies to the Chinese *Tien* and the Mongolian *Tengri*. And this is not all. By a fortunate accident the Turanian name of *Tengri* can be traced back from the modern Mongolian to an earlier period. Chinese authors, in their accounts of the early history of the Huns, tell us that the title given by the Huns to their leaders was *tangli-kutu* (or tchen-jü).[2] This *tangli-kutu* meant in their language Son of Heaven, and you will remember that the same name, Son of Heaven, is still given to the Chinese Emperor.[3] It does not mean Son of God, but Emperor by the grace of God. Now the Chinese title is *tien-tze*, corresponding to the Hunnish *tangli-kutu*. Hence Hunnish *tang-li*, or Mongolian *teng-ri*, are the same as the Chinese *Tien*.

Again, in the historical accounts which the Chinese give of the *Tukiu*, the ancestors of the Turks, it is said that they worshipped the spirits of the Earth, and that they called these spirits *pu-teng-i-li*. Here the first

---

[1] Turkish "tangry" (or tenri), the Yakute "tangara."

[2] Schott, *Ueber dans Altaische Sprachgeschlecht*, p. 9.

[3] Schott, *Chinesische Literatur*, p. 60.

syllable must be intended for earth, while in *teng-i-li* we have again the same word as the Mongolian *tengri*, only used, even at that early time, no longer in the sense of heaven, or god of heaven, but as a name of gods and spirits in general. We find a similar transition of meaning in the modern Yakute word *tangara*. It means the sky, and it means God; but among the Christian converts in Siberia, *tangara* is also used to signify " the Saints." The wild reindeer is called in Yakute God's reindeer, because it lives in the open air, or because God alone takes care of it.

Here, then, we have the same kind of evidence which enabled us to establish a primitive Aryan and a primitive Semitic religion: we have a common name, and this name given to the highest deity, preserved in the monosyllabic language of China, and in the cognate, though agglutinative, dialects of some of the principal North Turanian tribes. We find in these words, not merely a vague similarity of sound and meaning, but, watching their growth in Chinese, Mongolian, and Turkish, we are able to discover in them traces of organic identity. Everywhere they begin with the meaning of sky, they rise to the meaning of God, and they sink down again to the meaning of gods and spirits. The changes in the meaning of these words run parallel with the changes that took place in the religions of these nations, which in China, as elsewhere, combine the worship of numberless spirits with the belief in a supreme heavenly deity.

Did we allow ourselves to be guided by mere similarity of sound and meaning, it would be easy, for instance, to connect the name given to the highest deity by the Samoyedes, *Num*, the same as the Finnish *Juma*(*la*), with the name used for God in the language of Tibet, *Nam*. This might seem a most important link, because it would help us to establish an original identity of religion among members of the North and South Turanian branches. But till we know something of the antecedents of the Tibetan word, till we know, as I said before, its organic growth, we cannot think of using it for such purposes.

If we now turn for a moment to the minor spirits believed in by the large masses in China, we shall easily see that they, too, in their character are strikingly like the spirits worshipped by the North Turanian tribes. These spirits in Chinese are called *Shin*,[1] which is really the name given to every invisible power or influence which can be perceived in operation in the universe. Some *Shins* or spirits receive real worship, which is graduated according to their dignity; others are looked upon with fear. The spirits of pestilence are driven out and dispersed by exorcism; many are only talked about. There are so many spirits that it seems impossible to fix their exact number. The principal classes [2] are the celestial spirits (tien shin), the terrestrial spirits (ti ki), and the ancestral spirits (jin kwei), and this is the order [3] in which they are ranked according to their dignity. Among celestial spirits (tien shin) we find the spirits of the sun and the moon and the stars, the clouds, wind, thunder, and rain; among terrestrial spirits, those of the mountains, the fields, the grain, the rivers, the trees, the year. Among the departed spirits are those of the emperors, the sages, and other public benefactors, which are to be re-

---

[1] Medhurst, *Reply*, p. 11.

[2] Medhurst, *Reply*, l. c., p. 21.

[3] Ibid. l. c., p. 22. The spirits of heaven are called *shin;* the spirits of earth are called *ki;* when men die their wandering and transformed souls and spirits are called *kwei.*

vered by the whole nation, while each family has its own *manes* which are treated with special reverence and honored by many superstitious rites.[1]

The same state of religious feeling is exhibited among the North Turanian tribes, only without those minute distinctions and regulations in which the Chinese mind delights. The Samoyedes, as we saw, believed in a supreme god of heaven, called *Num;* but Castrén, who lived so long among them, says: "The chief deities invoked by their priests or sorcerers, the Shamans, are the so-called *Tadebejos*,[2] invisible spirits dwelling in the air, the earth, the water, and everywhere in nature. I have heard many a Samoyede say that they were merely the spirits of the departed, but others look upon them as a class of inferior deities."

The same scholar tells us [3] that "the mythology of the Finns is flooded with names of deities. Every object in nature has a genius, called *haltia*, which is supposed to be its creator and protector. These spirits were not tied to these outward objects, but were free to roam about, and had a body and soul, and their own well-marked personality. Nor did their existence depend on the existence of a single object; for though there was no object in nature without a genius, the genius was not confined to any single object, but comprehended the whole class or genus. This mountain-ash, this stone, this house has its own genius, but the same genius cares for all other mountain-ashes, stones, and houses."

[1] Medhurst, *Reply,* i. p. 43. The great sacrifices are offered only to *Te* or *Shang-te,* the same as *Tien.* The five *Te* which used to be joined with *Shang-te* at the great border sacrifice were only the five powers or qualities of *Shang-te* personified. Since the year A. D. 1369, the worship of these five *Te* has been abolished.

[2] Castren, *Finnische Mythologie,* p. 122.   [3] Page 105.

We have only to translate this into the language of logic, and we shall understand at once what has happened here as elsewhere in the growth of religious ideas and mythological names. What we call a general conception, or what used to be called "*essentia generalis*," "the tree-hood," "the stone-hood," "the house-hood," in fact, the genus tree, stone, and house, is what the Finns and Samoyedes call the genius, the *haltia*, the *tadebejo*, and what the Chinese call *Shin*. We speak very glibly of an *essentia generalis*, but to the unschooled mind this was too great an effort. Something substantial and individual had to be retained when trees had to be spoken of as a forest, or days as a year; and in this transition period from individual to general conceptions, from the tangible to the comprehensible, from the real to the abstract, the shadow, the ghost, the power or the spirit of the forest, of the year, of the clouds, and the lightning, took possession of the human mind, and a class of beings was called into existence which stands before us as so-called deities in the religion and mythology of the ancient world.

The worship of ancestral spirits is likewise shared in common by the North Turanian races and the Chinese. I do not lay much stress on that fact, because the worship of the spirits of the departed is perhaps the most widely spread form of natural superstition all over the world. It is important, however, to observe that on this point also, which has always been regarded as most characteristic of Chinese religion, there is no difference between China and Northern Asia. Most of the Finnish and Altaic tribes, says Castrén,[1] cherish a belief that death, which they look upon with terrible fear, does not entirely destroy individual existence. And even those

[1] Page 119.

who do not profess belief in a future life, observe certain ceremonies which show that they think of the departed as still existing. They take food, dresses, oxen, knives, tinder-boxes, kettles, and sledges, and place them on the graves; nay, if pressed, they would confess that this is done to enable the departed to hunt, to fish, and to fight, as they used to do when alive. Lapps and Finns admit that the body decays, but they imagine that a new body is given to the dead in the lower world. Others speak of the departed as ghosts or spirits, who either stay in the grave or in the realms of the dead, or who roam about on earth, particularly in the dead of night, and during storm and rain. They give signs of themselves in the howling of the wind, the rustling of leaves, the crackling of the fire, and in a thousand other ways. They are invisible to ordinary mortals, but the sorcerers or Shamans can see them, and can even divine their thoughts. It is curious that in general these spirits are supposed to be mischievous; and the most mischievous of all are the spirits of the departed priests.[1] They interrupt the sleep, they send illness and misfortunes, and they trouble the conscience of their relatives. Everything is done to keep them away. When the corpse has been carried out of the house, a red hot stone is thrown after the departed, as a charm to prevent his return. The offerings of food and other articles deposited on the grave are accounted for by some as depriving the dead of any excuse for coming to the house, and fetching these things himself. Among the Tchuvashes a son uses the following invocation when offering sacrifice to the spirit of his father: "We honor thee with a feast; look, here is bread for thee, and different kinds of meat; thou hast all thou

[1] Page 123.

canst want; but do not trouble us, do not come near us."[1]

It is certainly a general belief that if they receive no such offerings, the dead revenge themselves by sending diseases and other misfortunes. The ancient Hiongnu or Huns killed the prisoners of war on the tombs of their leaders; for the Shamans assured them that the anger of the spirits could not be appeased otherwise. The same Huns had regular sacrifices in honor of their ancestral spirits. One tribe, the Topas, which had migrated from Siberia to Central Asia, sent ambassadors with offerings to the tombs of their ancestors. Their tombs were protected with high palings, to prevent the living from clambering in, and the dead from clambering out. Some of these tombs were magnificently adorned,[2] and at last grew almost, and in China[3] altogether, into temples where the spirits of the departed were actually worshipped. All this takes place by slow degrees; it begins with placing a flower on the tombs; it ends with worshipping the spirits of departed emperors[4] as equals of the Supreme Spirit, the *Shang-te* or *Tien*, and as enjoying a divine rank far above other spirits or *Shins*. The difference at first sight, between the minute ceremonial of China and the homely worship of Finns and Lapps may seem enormous; but if we trace both back as far as we can, we see that the early stages of their religious belief are curiously alike. First, a worship of heaven, as the emblem of the most exalted conception which the un-

---

[1] Page 122. [2] Castrén, l. c., p. 122.

[3] When an emperor died, and men erected an ancestral temple, and set up a parental tablet (as a resting-place for the "shin" or spirit of the departed), they called him Te. — Medhurst, *Inquiry*, p. 7; from *Le-ke*, vol. i. p. 49.

[4] Medhurst, *Inquiry*, p. 45.

tutored mind of man can entertain, expanding with the expanding thoughts of its worshippers, and eventually leading and lifting the soul from horizon to horizon, to a belief in that which is beyond all horizons, a belief in that which is infinite. Secondly, a belief in deathless spirits or powers of nature; which supplies the more immediate and every-day wants of the religious instinct of man, satisfies the imagination, and furnishes the earliest poetry with elevated themes. Lastly, a belief in the existence of ancestral spirits; which implies, consciously or unconsciously, in a spiritual or in a material form, that which is one of the life-springs of all religion, a belief in immortality.

Allow me in conclusion to recapitulate shortly the results of this lecture.

We found, first of all, that there is a natural connection between language and religion, and that therefore the classification of languages is applicable to the ancient religions of the world.

We found, secondly, that there was a common Aryan religion before the separation of the Aryan race; a common Semitic religion before the separation of the Semitic race; and a common Turanic religion before the separation of the Chinese and the other tribes belonging to the Turanian class. We found, in fact, three ancient centres of religion as we had found before three ancient centres of language, and we have thus gained, I believe, a truly historical basis for a scientific classification of the principal religions of the world.

FOURTH LECTURE.

WHEN I came to deliver the first of this short course of lectures, I confess I felt sorry for having undertaken so difficult a task; and if I could have withdrawn from it with honor, I should gladly have done so. Now that I have only this one lecture left, I feel equally sorry, and I wish I could continue my course, in order to say something more of what I wished to say, and what in four lectures I could say but very imperfectly. From the announcement of my lectures you must have seen that in what I called "An Introduction to the Science of Religion" I did not intend to treat of more than some preliminary questions. I chiefly wanted to show in what sense a truly scientific study of religion was possible, what materials there are to enable us to gain a trustworthy knowledge of the principal religions of the world, and according to what principles these religions may be classified. It would perhaps have been more interesting to some of my hearers if we had rushed at once into the ancient temples to look at the broken idols of the past, and to discover, if possible, some of the fundamental ideas that found expression in the ancient systems of faith and worship. But in order to explore with real advantage any ruins, whether of stone or of thought, it is necessary that we should know where to look and how to look. In most works on the history of ancient religions we are driven about like forlorn tourists in a vast museum where ancient and modern statues, gems of Oriental and European workmanship, original works of art and mere copies are piled up together, and at the end of our journey we only feel bewildered and disheartened. We have seen much, no doubt, but we carry away very little. It is better, before we enter into these labyrinths, that we should spend a few hours in making up our minds as to what we really want to see and what we may pass by; and if in these introductory lectures we have arrived at a clear view on these points, you will find hereafter that our time has not been spent in vain.

Throughout these introductory lectures, you will have observed that I have carefully abstained from entering on the domain of what I call *Theoretic*, as distinguished from *Comparative Theology*. Theoretic theology, or, as it is called, the philosophy of religion, has, as far as I can judge, its right place at the end, not at the beginning of comparative theology. I make no secret of my own conviction that a study of comparative theology will produce with regard to theoretic theology the same revolution which a study of comparative philology has produced in what used to be called the philosophy of language. You know how all speculations on the nature of language, on its origin, its development, its natural growth and inevitable decay have had to be taken up afresh from the very beginning, after the new light thrown on the history of language by the comparative method. I look forward to the same results with respect to philosophical inquiries into the nature of religion, its origin, and its development. I do not mean to say that all former speculations on these subjects will become useless. Plato's "Cratylus," even the "Hermes" of Harris, and Horne Tooke's "Diversions of Purley" have not become useless after the work done by Grimm and Bopp, by Humboldt and Bunsen. But I believe that philosophers who speculate on the origin of religion and on the psychological conditions of faith, will in future write more circumspectly, and with less of that dogmatic assurance which has hitherto distinguished so many speculations on the philosophy of religion, not excepting those of Schelling and Hegel. Before the rise of geology it was easy to speculate on the origin of the earth; before the rise of glossology, any theories on the revealed, the mimetic, the interjectional, or the conventional origin of language might easily be held and defended. Not so now, when facts have filled the place that was formerly open to theories, and when those who have worked most carefully among the *débris* of the earth or the strata of languages are most reluctant to approach the great problem of the first beginnings.

So much in order to explain why in this introductory course I have confined myself within narrower limits than some of my hearers seem to have expected. And now, as I have but one hour left, I shall try to make the best use of it I can, by devoting it entirely to the point on which I have not yet touched, namely, on the right spirit in which ancient religions ought to be studied and interpreted.

No judge, if he had before him the worst of criminals, would treat him as most historians and theologians have treated the religions of the world. Every act in the lives of their founders, which shows that they were but men, is eagerly seized and judged without mercy; every doctrine that is not carefully guarded is interpreted in the worst sense that it will bear; every act of worship that differs from our own way of serving God is held up to ridicule and contempt. And this is not done by accident, but with a set purpose, nay, with something of that artificial sense of duty which stimulates the counsel for the defense to see nothing but an angel in one client, and anything but an angel in the plaintiff on the other side. The result has been — as it could not be otherwise — a complete miscarriage of justice, an utter misapprehension of the real character and purpose of the ancient religions of mankind; and, as a necessary consequence, a failure in discovering the peculiar features which really distinguish Christianity from all the religions of the world, and secure to its founder his own peculiar place in the history of the world, far away from Vasish*tha*, Zoroaster, and Buddha, from Moses and Mohammed, from Confucius and Lao-tse. By unduly depreciating all other religions, we have placed our own in a position which its founder never intended for it; we have torn it away from the sacred context of the history of the world; we have ignored, or willfully narrowed, the sundry times and divers manners in which, in times past, God spake unto the fathers by the prophets; and instead of recognizing Christianity as coming in the fullness of time, and as the fulfillment of the hopes and desires of the whole world, we have brought ourselves to look upon its advent as the only broken link in that unbroken chain which is rightly called the Divine government of the world. Nay, worse than this: there are people who, from mere ignorance of the ancient religions of mankind, have adopted a doctrine more unchristian than any that could be found in the pages of the religious books of antiquity, namely, that all the nations of the earth, before the rise of Christianity, were mere outcasts, forsaken and forgotten of their Father in heaven, without a knowledge of God, without a hope of salvation. If a comparative study of the religions of the world produced but this one result, that it drove this godless heresy out of every Christian heart, and made us see again in the whole history of the world the eternal wisdom and love of God toward all his creatures, it would have done a good work. And it is high time that this good work should be done. We have learnt to do justice to the ancient poetry, the political institutions, the legal enactments, the systems of philosophy, and the works of art of nations differing from ourselves in many respects; we have brought ourselves to value even the crude and imperfect beginnings in all these spheres of mental activity; and I believe we have thus learnt lessons from ancient history which we could not have learnt anywhere else. We can admire the temples of the ancient world, whether in Egypt, Babylon, or Greece; we can stand in raptures before the statues of Phidias; and only when we approach the religious conceptions which find their expression in the temples of Minerva and in the statues of Jupiter, we turn away with pity or scorn, we call their gods mere idols and images, and class their worshippers — Perikles, Phidias, Sokrates, and Plato — with the worshippers of stocks and stones. I do not deny that the religions of the Babylonians, Egyptians, Greeks, and Romans were imperfect and full of errors, particularly in their later stages; but I maintain that the fact of these ancient people having any religion at all, however imperfect, raises them higher, and brings them nearer to us, than all their works of art, all their poetry, all their philosophy. Neither their art nor their poetry nor their philosophy would have been possible without religion; and if we will but look without prejudice, if we will but judge as we ought always to judge, with unwearying love and charity, we shall be surprised at that new world of beauty and truth which, like the azure of a vernal sky, rises before us from behind the clouds of the ancient mythologies.

We can speak freely and fearlessly; *we* can afford to be charitable. There was a time when it was otherwise. There was a time when people imagined that truth, particularly the highest truth, the truth of religion, could only conquer by blind zeal, by fire and sword. At that time all idols were to be overthrown, their altars to be destroyed, and their worshippers to be cut to pieces. But there came a time when the sword was to be put up into its place. . . . . And if after that time there was a work to work and a fight to fight, which required the fiery zeal of apostles and martyrs, that time also is now past; the conquest is gained, and we have time to reflect calmly on what is past and what is still to come. We are no longer afraid of Baal or Jupiter. Our dangers and our difficulties are now of a very different kind. If we believe that there is a God, and that He created heaven and earth, and that He ruleth the world by his unceasing providence, we cannot believe that millions of human beings, all created like ourselves in the image of God, were, in their time of ignorance, so utterly abandoned that their whole religion was falsehood, their whole worship a farce, their whole life a mockery. An honest and independent study of the religions of the world will teach us that it was not so, — will teach us the same lesson which it taught St. Augustine,— that there is no religion which does not contain some grains of truth. Nay, it will teach us more: it will enable us to see in the history of the ancient religions, more clearly than anywhere else, the *Divine education of the human race.*

I know this is a view which has been much objected to, but I hold it as strongly as ever. If we must not read in the history of the whole human race the daily lessons of a Divine teacher and guide, if there is no purpose, no increasing purpose in the succession of the religions of the world, then we might as well shut up the godless book of history altogether, and look upon men as no better than the grass which is to-day in the field and to-morrow cast into the oven. Man would then be indeed of less value than the sparrows, for none of them is forgotten before God. But those who imagine that, in order to make sure of their own salvation, they must have a great gulf fixed between themselves and all the other nations of the world, — between their own religion and the religions of Zoroaster, Buddha, or Confucius, — can hardly be aware how strongly the interpretation of the history of the religions of the world, as an education of the human race, can be supported by authorities before which they themselves would probably bow in silence. We need not appeal to a living bishop to prove the soundness, or to a German philosopher to prove the truth of this view. If we wanted authorities we could appeal to Popes, to the Fathers of the Church, to the Apostles themselves, for they have all upheld the same view with no uncertain voice.

I pointed out before that the simultaneous study of the Old and the New Testament, with an occasional reference to the religion and philosophy of Greece and Rome, had supplied Christian divines with some of the most useful lessons for a wider comparison of all the religions of the world. In studying the Old Testament, and observing in it the absence of some of the most essential truths of Christianity, they, too, had asked with surprise why the interval between the fall of man and his redemption had been so long, why men were allowed so long to walk in darkness, and whether the heathens had really no place in the counsels of God. Here is the answer of a Pope, of Leo the Great [1] (440–461) : —

"Let those who with impious murmurings find fault with the Divine dispensations, and who complain about the lateness of Our Lord's nativity, cease from their grievances, as if what was carried out in this last age of the world had not been impending in time past. .... What the apostles preached, the prophets had announced before, and what has always been believed cannot be said to have been fulfilled too late. By this delay of his work of salvation the wisdom and love of God have only made us more fitted for his call; so that, what had been announced before by many signs and words and mysteries during so many centuries, should not be doubtful or uncertain in the days of the gospel. .... God has not provided for the interests of men by a new counsel or by a late compassion; but He had instituted from the beginning for all men one and the same path of salvation."

This is the language of a Pope — of Leo the Great. Now let us hear what St. Irenæus says, and how he explains to himself the necessary imperfection of the early religions of mankind. "A mother," he says, "may indeed offer to her infant a complete repast, but her infant cannot yet receive the food which is meant for full-grown men. In the same manner God might indeed from the beginning have offered to man the

---

[1] Hardwick, *Christ and other Masters*, vol. i. p. 85.

truth in its completeness, but man was unable to receive it, for he was still a child."

If this, too, is considered a presumptuous reading of the counsels of God, we have, as a last appeal, the words of St. Paul, that "the law was the school-master to the Jews," joined with the words of St. Peter, "Of a truth I perceive that God is no respecter of persons, but in every nation he that feareth Him and worketh righteousness is accepted with Him."

But, as I said before, we need not appeal to any authorities if we will but read the records of the ancient religions of the world with an open heart and in a charitable spirit — in a spirit that thinketh no evil, but rejoiceth in the truth wherever it can be found.

I suppose that most of us, sooner or later in life, have felt how the whole world — this wicked world, as we call it — is changed as if by magic, if once we can make up our mind to give men credit for good motives, never to be suspicious, never to think evil, never to think ourselves better than our neighbors. Trust a man to be true and good, and, even if he is not, your trust will tend to make him true and good. It is the same with the religions of the world. Let us but once make up our mind to look in them for what is true and good, and we shall hardly know our old religion again. If they are the work of the devil, as many of us have been brought up to believe, then never was there a kingdom so divided against itself from the very beginning. There is no religion — or, if there is, I do not know it — which does not say, "Do good, avoid evil." There is none which does not contain what Rabbi Hillel called the quintessence of all religions, the simple warning, "Be good, my boy." "Be good, my boy," may seem a very short catechism; but let us add to it,

"Be good, my boy, for God's sake," and we have in it very nearly the whole of the Law and the Prophets.

I wish I could read you the extracts I have collected from the sacred books of the ancient world, grains of truth more precious to me than grains of gold; prayers so simple and so true that we could all join in them if we once accustomed ourselves to the strange sounds of Sanskrit or Chinese. I can to-day give you a few specimens only.

Here is a prayer of Vasish*th*a, a Vedic prophet, addressed to Varu*n*a, the Greek Οὐρανὸς, an ancient name of the sky and of the god who resides in the sky.

I shall read you one verse at least in the original — it is the 86th hymn of the seventh book of the Rig-Veda — so that you may hear the very sounds which more than three thousand years ago were uttered for the first time in a village on the borders of the Sutledge, then called the *S*atadru, by a man who felt as we feel, who spoke as we speak, who believed in many points as we believe — a dark-complexioned Hindu, shepherd, poet, priest, patriarch, and certainly a man who, in the noble army of prophets, deserves a place by the side of David. And does it not show the indestructibility of the spirit, if we see how the waves which, by a poetic impulse, he started on the vast ocean of thought have been heaving, and spreading, and widening, till after centuries and centuries they strike against our shores and tell us, in accents that cannot be mistaken, what passed through the mind of that ancient Aryan poet when he felt the presence of an Almighty God, the maker of heaven and earth, and felt at the same time the burden of his sin, and prayed to his God that He might take that burden from him, that He might forgive him his sin. When you listen to the strange sounds of the Vedic hymn, you are listening, even in this Royal Institution, to spirit-rapping — to real spirit-rapping. Vasishtha is really among us again, and if you will accept me as interpreter, you will find that we can all understand what the old poet wished to say: —

> Dhira tv asya mahina ganumshi,
> vi yas tastambha rodasi kid urvi,
> pra nakam rishvam nunude brihantam,
> dvita nakshatram paprathak ka bhuma.

Wise and mighty are the works of him who stemmed asunder the wide firmaments (heaven and earth). He lifted on high the bright and glorious heaven; he stretched out apart the starry sky and the earth.

Do I say this to my own self? How can I get near unto Varuna? Will he accept my offering without displeasure? When shall I, with a quiet mind, see him propitiated?

I ask, O Varuna, wishing to know this my sin; I go to ask the wise. The sages all tell me the same: "Varuna it is who is angry with thee."

Was it for an old sin, O Varuna, that thou wishest to destroy thy friend, who always praises thee? Tell me, thou unconquerable Lord! and I will quickly turn to thee with praise, freed from sin.

Absolve us from the sins of our fathers, and from those which we committed with our own bodies. Release Vasishtha, O King, like a thief who has feasted on stolen cattle; release him like a calf from the rope.

It was not our own doing, O Varuna, it was a slip; an intoxicating draught, passion, dice, thoughtlessness. The old is there to mislead the young; even sleep is not free from mischief.

Let me without sin give satisfaction to the angry god, like a slave to his bounteous lord. The lord god enlightened the foolish; he, the wisest, leads his worshipper to wealth.

O lord Varuna, may this song go well to thy heart! May we prosper in keeping and acquiring! Protect us, O gods, always with your blessings.

I am not blind to the blemishes of this ancient

prayer, but I am not blind to its beauty either, and I think you will admit that the discovery of even one such poem among the hymns of the Rig-Veda, and the certainty that such a poem was composed in India at least three thousand years ago, without any inspiration but that which all can find who seek for it if haply they may find it, is well worth the labor of a life. It shows that man was never forsaken of God, and that conviction is worth more to the student of history than all the dynasties of Babylon and Egypt, worth more than all lacustrine villages, worth more than the skulls and jaw-bones of Neanderthal or Abbeville.

My next extract will be from the Zendavesta, the sacred book of the Zoroastrians, older in its language than the cuneiform inscriptions of Cyrus, Darius, Xerxes, and still believed in by a small remnant of the Persian race, now settled at Bombay, and known all over the world by the name of Parsis.[1]

I ask thee, tell me the truth, O Ahura! Who was from the beginning the father of the pure creatures? Who has made a path for the sun and for the stars? Who (but thou) makes the moon to increase and to decrease? That, O Mazda, and other things, I wish to know.

I ask thee, tell me the truth, O Ahura! Who holds the earth and the clouds that they do not fall? Who holds the sea and the trees? Who has given swiftness to the wind and the clouds? Who is the creator of the good spirit?

I ask thee, tell me the truth, O Ahura! Who has made the kindly light and the darkness, who has made the kindly sleep and the awaking? Who has made the mornings, the noons, and the nights? Who has made him who ponders on the measure of the laws?

We cannot always be certain that we have found the right sense of the Zendavesta, for its language is

[1] *Yasna*, xliv. 3, ed. Brockhaus, p. 130; Spiegel, *Yasna*, p. 146; Haug, *Essays*, p. 150.

full of difficulties; yet so much is clear, that in the Bible of Zoroaster every man is called upon to take his part in the great battle between Good and Evil which is always going on, and is assured that in the end good will prevail.

What shall I quote from Buddha? for there is so much in his sayings and his parables that it is indeed difficult to choose. In a collection of his sayings, written in Pâli,— of which I have lately published a translation,[1] — we read : —

1. All that we are is the result of what we have thought ; it is founded on our thoughts, it is made up of our thoughts. If a man speaks or acts with an evil thought, pain follows him as the wheel follows the foot of him who draws the cart.

49. As the bee collects honey and departs without injuring the flower, so let the sage dwell on earth.

62. "These sons belong to me, and this wealth belongs to me," with such thoughts a fool is tormented. He himself does not belong to himself; how much less sons and wealth!

121. Let no man think lightly of evil, saying in his heart, It will not come nigh unto me. Let no man think lightly of good, saying in his heart, It will not benefit me. Even by the falling of water-drops a water-pot is filled.

173. He whose evil deeds are covered by good deeds, brightens up this world like the moon when she rises from behind the clouds.

223. Let a man overcome anger by love, evil by good, the greedy by liberality, the liar by truth.

264. Not by tonsure does an undisciplined man become a saint : can a man be a saint who is still held captive by desires and greediness ?

394. What is the use of platted hair, O fool! what of the raiment of goat-skins ? Within thee there is ravening, but the outside thou makest clean.

[1] *Buddhaghosha's Parables*, translated from Burmese by Captain Rogers ; with an Introduction containing Buddha's "Dhammapada," or "Path of Virtue," translated from Pâli by Max Müller. London : Trübner & Co., 1870.

In no religion are we so constantly reminded of our own as in Buddhism, and yet in no religion has man been drawn away so far from the truth as in the religion of Buddha. Buddhism and Christianity are indeed the two opposite poles with regard to the most essential points of religion; *Buddhism* ignoring all feeling of dependence on a higher power, and therefore denying the very existence of a supreme Deity; *Christianity* resting entirely on a belief in God as the Father, in the Son of Man as the Son of God, and making us all children of God by faith in his Son. Yet between the language of Buddha and his disciples and the language of Christ and his apostles there are strange coincidences. Even some of the Buddhist legends and parables sound as if taken from the New Testament, though we know that many of them existed before the beginning of the Christian era.

Thus, one day Ananda, the disciple of Buddha, after a long walk in the country, meets with Matangi, a woman of the low caste of the *Kandalas*, near a well, and asks her for some water. She tells him what she is, and that she must not come near him. But he replies, " My sister, I ask not for thy caste or thy family, I ask only for a draught of water." She afterwards becomes herself a disciple of Buddha.[1]

While in the New Testament we read, " If thy right eye offend thee, pluck it out and cast it from thee," we find among the Buddhists a parable of a young priest whose bright and lovely eyes proved too attractive to a lady whom he visits, and who thereupon plucks out his right eye and shows it to her that she may see how hideous it is.[2]

[1] Burnouf, *Introduction à l'Histoire du Buddhisme*, p. 205.
[2] See *Katha-sarit-sagara*, ed. Brockhaus, vi. 28, p. 14.

8

According to Buddha, the motive of all our actions should be *pity* or *love* for our neighbor.

And as in Buddhism, so even in the writings of Confucius we find again what we value most in our religion. I shall quote but one saying of the Chinese sage.[1]

"What you do not like when done to yourself, do not do that to others."

One passage only from the founder of the second religion in China, from Lao-tse (cap. 25) : —

There is an Infinite Being, which existed before heaven and earth.
How calm it is! how free!
It lives alone, it changes not.
It moves everywhere, but it never suffers.
We may look on it as the Mother of the Universe.
I, I know not its name.
In order to give it a title, I call it *Tao* (the way).
When I try to give it a name, I call it *Great*.
After calling it *Great*, I call it *Fugitive*.
After calling it *Fugitive*, I call it *Distant*.
After calling it *Distant*, I say it comes back to me.

Need I say that Greek and Roman writers are full of the most exalted sentiments on religion and morality, in spite of their mythology and in spite of their idolatry? When Plato says that man ought to strive after likeness with God, do you think that he thought of Jupiter, or Mars, or Mercury? When another poet exclaimed that the conscience is a god for all men, was he so very far from a knowledge of the true God?

I wish we could explore together in this spirit the ancient religions of mankind, for I feel convinced that the more we know of them the more we shall see that

---

[1] Dr. Legge's *Life and Teachings of Confucius*, p. 47.

there is not one which is entirely false; nay, that in one sense every religion was a true religion, being the only religion which was possible at the time, which was compatible with the language, the thoughts, and the sentiments of each generation, which was appropriate to the age of the world. I know full well the objections that will be made to this. Was the worship of Moloch, it will be said, a true religion when they burnt their sons and their daughters in the fire to their gods? Was the worship of Mylitta, or is the worship of Kali a true religion, when within the sanctuary of their temple they committed abominations that must be nameless? Was the teaching of Buddha a true religion, when men were asked to believe that the highest reward of virtue and meditation consisted in a complete annihilation of the soul? Such arguments may tell in party warfare, though even there they have provoked fearful retaliation. Can that be a true religion, it has been answered, which consigned men of holy innocence to the flames, because they held that the Son was like unto the Father, but not the same as the Father, or because they would not worship the Virgin and the saints? Can that be a true religion which screened the same nameless crimes behind the sacred walls of monasteries? Can that be a true religion which taught the eternity of punishment without any hope of pardon or salvation for the sinner, however penitent? People who judge of religions in that spirit will never understand their real purport, will never reach their sacred springs. These are the excrescences, the inevitable excrescences of religion. We might as well judge of the health of a people from its hospitals, or of its morality from its prisons. If we want to judge of a religion, we must try to study it as much as possible in the mind of its founder; and when that is impossible, as it is but too often, we must try to find it in the lonely chamber and the sick-room, rather than in the colleges of augurs and the councils of priests.

If we do this, and if we bear in mind that religion must accommodate itself to the intellectual capacities of those whom it is to influence, we shall be surprised to find so much of true religion where we only expected degrading superstition or an absurd worship of idols.

The intention of religion, wherever we meet it, is always holy. However imperfect, however childish a religion may be, it always places the human soul in the presence of God; and however imperfect and however childish the conception of God may be, it always represents the highest ideal of perfection which the human soul, for the time being, can reach and grasp. Religion therefore places the human soul in the presence of its highest ideal, it lifts it above the level of ordinary goodness, and produces at least a yearning after a higher and better life — a life in the light of God. The expression that is given to these early manifestations of religious sentiment is no doubt frequently childish: it may be irreverent or even repulsive. But has not every father to learn the lesson of a charitable interpretation in watching the first stammerings of religion in his children? Why, then, should people find it so difficult to learn the same lesson in the ancient history of the world, and to judge in the same spirit the religious utterances of the childhood of the human race? Who does not recollect the startling and seemingly irreverent questionings of children about God, and who does not know how perfectly guiltless the child's mind is of real irreverence? Such outbursts of infantine religion hardly bear repeating. I shall only mention one instance. I well recollect the dismay which was created by a child exclaiming: "O! I wish there was at least *one* room in the house where I could play alone, and where God could not see me!" People who heard it were shocked; but to my mind, I confess, this childish exclamation sounded more wonderful than even the Psalm of David, "Whither shall I go from thy Spirit? or whither shall I flee from thy presence?"

It is the same with the childish language of ancient religion. We say calmly that God is omniscient and omnipresent. Hesiod speaks of the sun as the eye of Zeus that sees and perceives everything. Aratus wrote, "Full of Zeus are all the streets, all the markets of men; full of him is the sea and the harbors . . . . and we are also his offspring."

A Vedic poet, though of more modern date than the one I quoted before, speaking of the same Varuna whom Vasishtha invoked, says: "The great lord of these worlds sees as if he were near. If a man thinks he is walking by stealth, the gods know it all. If a man stands or walks or rides, if he goes to lie down or to get up, what two people sitting together whisper, King Varuna knows it, he is there as a third. This earth, too, belongs to Varuna, the king, and this wide sky with its ends far apart. The two seas (the sky and the ocean) are Varuna's loins; he is also contained in this small drop of water. He who should flee far beyond the sky, even he would not be rid of Varuna, the king. His spies proceed from heaven towards this world; with thousand eyes they overlook this earth. King Varuna sees all this, what is between heaven and earth, and what is beyond. He has counted the twinklings of our eyes. As a player throws down the dice, he settles all things."[1]

I do not deny that there is in this hymn much that is childish, that it contains expressions unworthy of the majesty of the Deity; but if I look at the language and the thoughts of the people who composed these hymns more than three thousand years ago, I wonder rather at the happy and pure expression which they have given to these deep thoughts than at the occasional harshnesses which jar upon our ears.

Ancient language is a difficult instrument to handle, particularly for religious purposes. It is impossible in human language to express abstract ideas except by metaphor, and it is not too much to say that the whole dictionary of ancient religion is made up of metaphors. With us these metaphors are all forgotten. We speak of spirit without thinking of breath, of heaven without thinking of the sky, of pardon without thinking of a release, of revelation without thinking of a veil. But in ancient language every one of these words, nay, every word that does not refer to sensuous objects, is still in a chrysalis stage: half material and half spiritual, and rising and falling in its character according to the varying capacities of speakers and hearers. Here is a constant source of misunderstandings, many of which have maintained their place in the religion and in the mythology of the ancient world. There are two distinct tendencies to be observed in the growth of ancient religion. There is, on the one side, the struggle of the mind against the material character of language, a constant attempt to strip words of their coarse covering, and fit them, by main force, for the

[1] *Chips from a German Workshop*, vol. i. p. 41; Atharva-veda, iv. 16.

purposes of abstract thought. But there is, on the other side, a constant relapse from the spiritual into the material, and, strange to say, a predilection for the material sense instead of the spiritual. This action and reaction has been going on in the language of religion from the earliest times, and it is at work even now.

It seems at first a fatal element in religion that it cannot escape from this flux and reflux of human thought, which is repeated at least once in every generation between father and son, between mother and daughter; but if we watch it more closely we shall find, I think, that this flux and reflux constitutes the very life of religion.

Place yourselves in the position of those who first are said to have worshipped the sky. We say that they worshipped the sky, or that the sky was their god; and in one sense that is true, but in a sense very different from that which is usually attached to such statements. If we use "god" in the sense which it has now, then to say that the sky was their god is to say what is simply impossible. We might as well say that with them Spirit meant nothing but air. Such a word as God, in our sense of the word — such a word even as *deus* and θεός in Latin and Greek, or *deva* in Sanskrit, which could be used as a general predicate — did not and could not exist at that early time in the history of thought and speech. If we want to understand ancient religion, we must first try to understand ancient language. Let us remember, then, that the first materials of language supply expression for such impressions only as are received through the senses.

If, therefore, there was a root meaning to burn, to be bright, to warm, such a root might supply a recognized name for the sun and for the sky. But let us now imagine, as well as we can, the process which went on in the human mind before the name of sky could be torn away from its material object and be used as the name of something totally different from the sky. There was in the heart of man, from the very first, a feeling of incompleteness, of weakness, of dependence, whatever we like to call it in our abstract language. We can explain it as little as we can explain why the new-born child feels the cravings of hunger and thirst. But it was so from the first, and is so even now. Man knows not whence he comes and whither he goes. He looks for a guide, for a friend; he wearies for some one on whom he can rest; he wants something like a father in heaven. In addition to all the impressions which he received from the outer world, there was in the heart of man a stronger impulse from within — a sigh, a yearning, a call for something that should not come and go like everything else, that should be before, and after, and forever, that should hold and support everything, that should make man feel at home in this strange world. Before this vague yearning could assume any definite shape it wanted a name; it could not be fully grasped or clearly conceived except by naming it. But where to look for a name? No doubt the store-house of language was there, but from every name that was tried the mind of man shrank back because it did not fit, because it seemed to fetter rather than to wing the thought that fluttered within and called for light and freedom. But when at last a name or even many names were tried and chosen, let us see what took place, as far as the mind of man was concerned. A certain satisfaction, no doubt, was gained by having a name or several names, however imperfect; but these names, like all other names, were but signs — poor, imperfect signs; they were predicates, and very partial predicates, of various small portions only of that vague and vast something which slumbered in the mind. When the name of the brilliant sky had been chosen, as it has been chosen at one time or other by nearly every nation upon earth, was sky the full expression of that within the mind which wanted expression? Was the mind satisfied? Had the sky been recognized as its god? Far from it. People knew perfectly well what they meant by the visible sky; the first man who, after looking everywhere for what he wanted, and who at last in sheer exhaustion grasped at the name of sky as better than nothing, knew but too well that his success was after all a miserable failure. The brilliant sky, was, no doubt, the most exalted; it was the only unchanging and infinite being that had received a name, and that could lend its name to that as yet unborn idea of the Infinite which disquieted the human mind. But let us only see this clearly, that the man who chose that name did not mean, could not have meant that the visible sky was all he wanted, that the blue canopy above was his god.

And now observe what happens when the name sky has thus been given and accepted. The seeking and finding of such a name, however imperfect, was the act of a manly mind, of a poet, of a prophet, of a patriarch, who could struggle, like another Jacob, with the idea of God that was within him, till he had found some name for it. But when that name had to be used with the young and the aged, with silly children and doting grandmothers, it was impossible to preserve it from being misunderstood. The first step downwards would be to look upon the sky as the abode of that being which was called by the same name; the next step would be to forget altogether what was behind the name, and to implore the sky, the visible canopy over our heads, to send rain, to protect the fields, the cattle, and the corn, to give to man his daily bread. Nay, very soon those who warned the world that it was not the visible sky that was meant, but that what was meant was something high above, deep below, far away from the blue firmament, would be looked upon either as dreamers whom no one could understand or as unbelievers who despised the sky, the great benefactor of the world. Lastly, many things that were true of the visible sky would be told of its divine namesake, and legends would spring up, destroying every trace of the deity that once was hidden beneath that ambiguous name.

I call this variety of acceptation, this misunderstanding, which is inevitable in ancient and also modern religion, the *dialectic growth and decay*, or, if you like, the *dialectic life of religion*, and we shall see again and again how important it is in enabling us to form a right estimate of religious language and thought. The dialectic shades in the language of religion are almost infinite; they explain the decay, but they also account for the life of religion. You may remember that Jacob Grimm, in one of his poetical moods, explained the origin of High and Low German, of Sanskrit and Prakrit, of Doric and Ionic, by looking upon the high dialects as originally the language of women and children. We can observe, I believe, the same parallel streams in the language of religion. There is a high and there is a low dialect; there is a broad and there is a narrow dialect; there are dialects for men and for children, for clergy and laity, for the noisy streets and for the still and lonely chamber. And as the child on growing up to manhood has to unlearn the language of the nursery, its religion, too, has to be translated from a feminine into a more masculine dialect. This does not take place without a struggle, and it is this constantly recurring struggle, this inextinguishable desire to recover itself, which keeps religion from utter stagnation. From first to last religion is oscillating between these two opposite poles, and it is only if the attraction of one of the two poles becomes too strong, that the healthy movement ceases, and stagnation and decay set in. If religion cannot accommodate itself on the one side to the capacity of children, or if on the other side it fails to satisfy the requirements of men, it has lost its vitality, and it becomes either mere superstition or mere philosophy.

If I have succeeded in expressing myself clearly, I think you will understand in what sense it may be said that there is truth in all religions, even in the lowest. The intention which led to the first utterance of a name like sky, used no longer in its material sense, but in a higher sense, was right. The spirit was willing, but language was weak. The mental process was not, as commonly supposed, an identification of the definite idea of deity with sky: such a process is hardly conceivable; it was, on the contrary, a first attempt at defining the indefinite impression of deity by a name that should approximately or metaphorically render at least one of its most prominent features. The first framer of that name of the deity, I repeat it again, could as little have thought of the material heaven as we do when we speak of the kingdom of heaven.[1]

[1] Medhurst, *Inquiry*, p. 20.

And now let us observe another feature of ancient religion that has often been so startling, but which, if we only remember what is the nature of ancient language, becomes likewise perfectly intelligible. It is well known that ancient languages are particularly rich in synonyms, or, to speak more correctly, that in them the same object is called by many names — is, in fact, *polyonomous*. While in modern languages most objects have one name only, we find in ancient Sanskrit, in ancient Greek and Arabic, a large choice of words for the same object. This is perfectly natural. Each name could express one side only of the object that had to be named, and, not satisfied with one partial name, the early framers of language produced one name after the other, and after a time retained those which seemed most useful for special purposes. Thus, the sky might be called not only the brilliant, but the dark, the covering, the thundering, the rain-giving. This is the *polyonomy* of language, and it is what we are accustomed to call *polytheism* in religion. Aristotle said: "God, though He is one, has many names (is polyonomous) because He is called according to states into which He always enters anew."[1] The same mental yearning which found its first satisfaction in using the name of the brilliant sky as an indication of the Divine, would soon grasp at other names of the sky not expressive of brilliancy, and therefore more appropriate to a religious mood in which the Divine was conceived as dark, awful, all-powerful. Thus we find in Sanskrit, by the side of Dyaus, another name of the covering sky, Varuna, originally only another attempt at naming the Divine, but soon assuming a separate and independent existence.

[1] Arist. *De Mundo*, cap. vii. init.

But this is not all. The very imperfection of every name that had been chosen, their very inadequacy to express the fullness and infinity of the Divine, would keep up the search for new names till at last every part of nature in which an approach to the Divine could be discovered was chosen as a name of the Omnipresent. If the presence of the Divine was perceived in the strong wind, the strong wind became its name: if its presence was perceived in the earthquake and the fire, the earthquake and the fire became its names. Do you still wonder at polytheism or at mythology? Why, they are inevitable. They are, if you like, a *parler enfantin* of religion. But the world had its childhood, and when it was a child it spoke as a child, it understood as a child, it thought as a child; and, I say again, in that it spoke as a child its language was true, in that it believed as a child its religion was true. The fault rests with us, if we insist on taking the language of children for the language of men, if we attempt to translate literally ancient into modern language, oriental into occidental speech, poetry into prose.

It is perfectly true that at present few interpreters, if any, would take such expressions as the head, the face, the mouth, the lips, the breath of Jehovah in a literal sense. But what does it mean, then, if we hear one of our most honest and learned theologians declare that he can no longer read from the altar the words of the Bible, " God spake these words and said?" If we can make allowance for mouth and lips and breath, we can surely make the same allowance for words and their utterance. The language of antiquity is the language of childhood: aye, and we ourselves, when we try to reach the Infinite and the Divine by means of more abstract terms, are but like children trying to place a ladder against the sky.

The *parler enfantin* in religion is not extinct; it never will be. Not only have some of the ancient childish religions been kept alive, as, for instance, the religion of India, which is to my mind like a half fossilized megatherion walking about in the broad daylight of the nineteenth century; but in our own religion and in the language of the New Testament there are many things which disclose their true meaning to those only who know what language is made of, who have not only ears to hear but a heart to understand the real meaning of parables.

What I maintain, then, is this, that as we put the most charitable interpretation on the utterances of children, we ought to put the same charitable interpretation on the apparent absurdities, the follies, the errors, nay, even the horrors of ancient religion. When we read of Belus, the supreme god of the Babylonians, cutting off his own head, that the blood flowing from it might be mixed with the dust out of which men were to be formed, this sounds horrible enough; but depend upon it what was originally intended by this myth was no more than this, that there is in man an element of Divine life: that we are also his offspring. The same idea existed in the ancient religion of the Egyptians, for we read, in the 17th chapter of their "Ritual," that the Sun mutilated himself, and that from the stream of his blood he created all beings.[1] And the author of Genesis, too, when he wishes to express the same idea, can only use the same human and symbolical language; he can only say that "God

[1] Vicomte de Rouge in *Annales de Philosophie Chretienne*, Nov. 1869, p. 332.

formed man from the dust of the ground, and breathed into his nostrils the breath of life."

If we have once learnt to be charitable in the interpretation of the language of other religions, we shall more easily learn to be charitable in the interpretation of the language of our own; we shall no longer try to force a literal interpretation on words and sentences in our sacred books, which, if interpreted literally must lose their original purport and their spiritual truth. In this way, I believe that a comparative study of the religions of the world will teach us many a useful lesson in the study of our own: that it will teach us, at all events, to be charitable both abroad and at home.

LECTURE

ON

BUDDHIST NIHILISM.

BY F. MAX. MÜLLER, M. A.

PROFESSOR OF COMPARATIVE PHILOLOGY IN THE UNIVERSITY
OF OXFORD; MEMBER OF THE FRENCH INSTITUTE, ETC.

DELIVERED BEFORE THE GENERAL MEETING OF THE AS-
SOCIATION OF GERMAN PHILOLOGISTS, AT KIEL,
28TH SEPTEMBER, 1869.

(*Translated from the German.*)

BUDDHIST NIHILISM.

I MAY be mistaken, but my belief is that the subject which I have chosen for my discourse cannot be regarded as alien to the general interests of this assembly.

Buddhism in its numerous varieties continues still the religion of the majority of mankind, and will therefore always occupy a very prominent place in a comparative study of the religions of the world. But the science of comparative theology, although the youngest branch on the tree of human knowledge, will, for an accurate and fruitful study of antiquity, soon become as indispensable as comparative philology. For how can we truly understand and properly appreciate a people, its literature, art, politics, morals, and philosophy, its entire conception of life, without having comprehended its religion, not only in its outer aspect, but in its innermost being, in its deepest far-reaching roots?

What our great poet once said almost prophetically of languages, may also be said of religions, — "*He who knows only one knows none.*" As the true knowledge of a language requires a knowledge of languages, thus a true knowledge of religion requires a knowledge of religions. And however bold the assertion may sound, that all the languages of mankind have an Oriental origin, true it is that all religions, like the suns, have risen from the East.

Here, therefore, in treating religions scientifically (those of the Aryan as well as those of the Semitic races) the Oriental scholar lawfully enters into the "plenum" of philology, if philology still is, as our President told us yesterday, what it once intended and wished to be, namely, the true Humanitas, which, like an emperor of yore, could say of itself, " humani nihil a me alienum puto."

Now it has been the peculiar fate of the religion of Buddha, that among all the so-called false or heathenish religions, it almost alone has been praised by all and everybody for its elevated, pure, and humanizing character. One hardly trusts one's eyes on seeing Catholic and Protestant missionaries vie with each other in their praises of the Buddha; and even the attention of those who are indifferent to all that concerns religion must be arrested for a moment, when they learn from statistical accounts that no religion, not even the Christian, has exercised so powerful an influence on the diminution of crime as the old simple doctrine of the Ascetic of Kapilavastu. Indeed no better authority can be brought forward in this respect than that of a still living Bishop of the Roman Catholic Church. In his interesting work on the life of Buddha, the author, the Bishop of Ramatha, the Apostolic Vicar of Ava and Pegu, speaks with so much candor of the merits of the Buddhist religion, that we are often at a loss which most to admire, his courage or his learning. Thus he says in one place:[1] "There are many moral precepts equally commanded and enforced in common by both creeds. It will not be deemed rash

[1] Page 494.

to assert that most of the moral truths, prescribed by the gospel, are to be met with in the Buddhistic scriptures." In another place Bishop Bigandet says :[1] "In reading the particulars of the life of the last Buddha Gaudama, it is impossible not to feel reminded of many circumstances relating to our Saviour's life, such as it has been sketched out by the Evangelists."

I might produce many even stronger testimonies in honor of Buddha and Buddhism, but the above suffice for my purpose.

But then, on the other hand, it appears as if people had only permitted themselves to be so liberal in their praise of Buddha and Buddhism, because they could, in the end, condemn a religion which, in spite of all its merits, culminated in Atheism and Nihilism. Thus we are told by Bishop Bigandet :[2] "It may be said in favor of Buddhism, that no philosophico-religious system has ever upheld, to an equal degree, the notions of a savior and deliverer, and the necessity of his mission, for procuring the salvation of man, in a Buddhist sense. The *role* of Buddha, from beginning to end, is that of a deliverer, who preaches a law designed to secure to man the deliverance from all the miseries he is laboring under. But by an inexplicable and deplorable eccentricity, the pretended savior, after having taught man the way to deliver himself from the tyranny of his passions, leads him, after all, into the bottomless gulf of a total annihilation."

This language may have a slightly episcopal tinge, yet we find the same judgment, in almost identical words, pronounced by the most eminent scholars who have written on Buddhism. The warm discussions on this subject, which have recently taken place at the

[1] Page 495. [2] Page viii.

Académie des Inscriptions et Belles-Lettres of Paris, are probably known to many of those who are here present; but better still, the work of the man whose place has not yet been filled, either in the French Academy, or on the Council Board of German Science — the work of Eugène Burnouf, the true founder of a scientific study of Buddhism. Burnouf, too, in his researches arrives at the same result, namely, that Buddhism, as known to us from its canonical books, in spite of its great qualities, ends in Atheism and Nihilism.

Now, as to Atheism, it cannot be denied that, if we call the old gods of the Veda — Indra, and Agni, and Yama — gods, Buddha was an Atheist. He does not believe in the divinity of these deities. What is noteworthy is that he does not by any means deny their bare existence, just as little as St. Augustine and other Fathers of the Church endeavored to sublimize, or entirely explain away the existence of the Olympian deities. The founder of Buddhism treats the old gods as superhuman beings, and promises the believers that they shall after death be reborn into the world of the gods, and shall enjoy divine bliss with the gods. Similarly he threatens the wicked that after death they shall meet with their punishment in the subterranean abodes and hells, where the Asuras, Sarpas, Nâgas, and other evil spirits dwell, beings whose existence was more firmly rooted in the popular belief and language, than that even the founder of a new religion could have dared to reason them away. But, although Buddha assigned to these mediatized gods and devils, palaces, gardens, and a court, — not second to their former ones, — he yet deprived them of all their sovereign rights. Although, according to Buddha, the worlds of the gods last for millions of years, they must perish, at the end of every Kalpa, with the gods and with the spirits who in the circle of births have raised themselves to the world of the gods. Indeed, the reorganization of the spirit-world goes further still. Already, before Buddha, the Brahmans had surmounted the low stand-point of mythological polytheism, and supplanting it by the idea of the Brahman, as the absolute divine or super-divine power. What, then, does Buddha decree? To this Brahman also he assigns a place in his universe. Over and above the world of the gods with its six paradises, he heaps up sixteen Brahma-worlds, not to be attained through virtue and piety only, but through inner contemplation, through knowledge and enlightenment. The dwellers in these worlds are already purely spiritualized beings, without body, without weight, without desire, far above men and gods. Indeed, the Buddhist architect rises to a still more towering height, heaping upon the Brahma-world four still higher worlds, which he calls the world of the formless. All these worlds are open to man, and the beings ascend and descend in the circle of time, according to the works they have performed, according to the truths they have recognized. But in all these worlds the law of change obtains; in none is there exemption from birth, age, and death. The world of the gods will perish like that of men, even the world of the formless will not last forever; but the Buddha, the Enlightened and truly Free, stands higher, and will not be affected or disturbed by the collapse of the Universe: "Si fractus illabatur orbis, impavidum ferient ruinæ."

Now, however, we meet with a vein of irony, which one would hardly have expected in Buddha. Gods and devils he had located; to all mythological and philosophical acquisitions of the past he had done justice as far as possible. Even fabulous beings, such as Nâgas, Gandharvas, and Garudas, had escaped the process of dissolution, which was to reach them later only at the hands of comparative mythology. There is only *one* idea, the idea of a personal creator, in regard to which Buddha is relentless.

It is not only denied, but even its origin, like that of an ancient myth, is carefully explained by him in its minutest details. This is done in the Brahmagâla-sûtra. Let us bear in mind that a destruction of the worlds occurs at the end of every kalpa, a destruction which not only annihilates earth and hell, but also all the worlds of the gods, and even the three lowest of the Brahma-worlds. A description of the duration of a kalpa can only be given in the language of Buddhism. Take a rock forming a cube of about fourteen miles, touch it once in a hundred years with a piece of fine cloth, and the rock will sooner be reduced to dust than a kalpa will have attained its end. It is said that at the end of the kalpa, after all the lower stories of the universe had been destroyed and a new world had again been slowly formed, the spirits dwelling in the higher Brahma-worlds had remained inviolate. Then one of these Spirits, a being without body, without weight, omnipresent and blessed within himself, descended, when his time had arrived, from the higher Brahma-world to the new-formed nether Brahma-world. There he first dwelt alone; but, by and by, the desire arose in him not to remain alone any longer. At the moment of the awakening of this desire within him, a second being accidentally descended from the higher into the lower Brahma-world. Then and there the thought originated in the first being, "I am the Brahma, the great Brahma, the Highest, the Unconquerable, the Omniscient, the Lord and King of All. I am the Creator of all things, the Father of All. *This* being has also been created by me; for as soon as I desired not to remain alone, my desire brought forth this second being." The other beings as they gradually descended from the higher worlds likewise believed that the first comer had been their Creator, for was he not older and mightier and handsomer than they?

But this is not all; for although it would explain how one spirit could consider himself the creator of other spirits, it would leave unexplained the circumstances of men on earth believing in such a creator. This is explained in the following manner: "In the course of time one of these higher beings sank lower and lower, and was finally born as a man on earth. There, by penances and deep meditation, he attained a state of inner enlightenment, which gives to man the faculty of remembering his former existences. He remembered the above narrated occurrences in the newly originated Brahma-world, and announced to mankind that there was a Creator, a Brahman, who had been prior to all other beings; that this Creator was eternal and immutable, while all beings created by him were mutable and mortal.

There is in this explanation, I believe, an unmistakable note of animosity, otherwise so alien to the character of Buddha, and the question naturally arises whether this can have been the doctrine of the founder of Buddhism himself. And herewith we at once approach our principal problem: "Is it possible to distinguish between Buddhism and the personal teaching of Buddha?" We possess the Buddhist canon and have a right to consider all that we find in this canon as orthodox Buddhist doctrine. But as there has been no lack of efforts in Christian theology to distinguish between the doctrine of the founder of our religion and that of the writers of the Gospels, to go beyond the canon of the New Testament, and to make the λόγια of the Master the only valid rule of our faith, so the same want was already felt at a very early period, among the followers of Buddha. King Asoka, the Indian Constantine, had to remind the assembled priests at the great council which had to settle the Buddhist canon, *that what had been said by Buddha that alone was well said.*[1] Works attributed to Buddha, but declared as apocryphal, or even as heterodox, already existed at that time.

Thus we are not by any means without an authority for distinguishing between Buddhism and the teaching of Buddha; the question is only whether such a separation is still practicable for us?

My belief is that all honest inquirers must oppose a No to this question. Burnouf never ventured to cast a glance beyond the boundaries of the Buddhist canon. What he finds in the canonical books, in the so-called " Three Baskets," is to him the doctrine of Buddha, similarly as we must accept, as the doctrine of Christ, what is contained in the four Gospels.

Still the question ought to be asked again, and again, whether, at least with regard to certain doctrines or facts, it may not be possible to make a step further in advance, even with the conviction that it cannot lead us to results of apodictic certainty. For if, as happens frequently, we find in the different parts of the canon,

---

[1] See Max Müller's *Chips from a German Workshop*, second edition, vol. i. p. xxiv.

views, not only differing from, but even contradictory to each other, it follows, I think, that one only of them can belong to Buddha personally, and I believe that in such a case we have the right to choose, and the liberty to accept *that* view as the original one, the one peculiar to Buddha, which *least* harmonizes with the later system of orthodox Buddhism.

As regards the denial of a Creator, or Atheism in the ordinary acceptation of the term, I do not think that any one passage from the books of the canon known to us, can be quoted which contradicts it, or which in any way presupposes the belief in a personal God or a Creator. All that may be urged are the words said to have been spoken by Buddha at the moment when he became the Enlightened, the Buddha. They are as follows: "Without ceasing shall I run through a course of many births, looking for the maker of this tabernacle, — and painful is birth again and again. But now, maker of the tabernacle, thou hast been seen; thou shalt not make up this tabernacle again. All thy rafters are broken, thy ridge-pole is sundered; the mind, being sundered, has attained to the extinction of all desires."

Here in the maker of the tabernacle, *i. e.* the body, one might be tempted to see a creator. But he who is acquainted with the general run of thought in Buddhism, soon finds that this architect of the house is only a poetical expression, and that whatever meaning may underlie it, it evidently signifies a force subordinated to the Buddha, the Enlightened.

But whilst we have no ground for exonerating the Buddha personally from the accusation of Atheism, the matter stands very differently as regards the charge of Nihilism. Buddhist Nihilism has always been much more incomprehensible than mere Atheism. A kind of religion is still conceivable, when there is something firm somewhere, when a something, eternal and self-dependent, is recognized, if not *without* and *above* man, at least *within* him. But if, as Buddhism teaches, the soul, after having passed through all the phases of existence, all the worlds of the gods and of the higher spirits, attains finally Nirvâna as its highest aim and last reward, *i. e.* becomes quite extinct, then religion is not any more what it ought to be — a bridge from the finite to the infinite, but a trap-bridge hurling man into the abyss, at the very moment when he thought he had arrived at the stronghold of the Eternal. According to the metaphysical doctrine of Buddhism, the soul cannot dissolve itself in a higher being, or be absorbed in the absolute substance, as was taught by the Brahmans and other mystics of ancient and modern times. For Buddhism knew not the Divine, the Eternal, the Absolute, and the soul, even as the I, or as the mere Self, the Atman, as called by the Brahmans, was represented in the orthodox Metaphysics of Buddhism as transient, as futile, as a mere phantom.

No person who reads with attention the metaphysical speculations on the Nirvâna contained in the Buddhist Canon, can arrive at any other conviction than that expressed by Burnouf, namely, that Nirvâna, the highest aim, the *summum bonum* of Buddhism, is the absolute nothing.

Burnouf adds, however, that this doctrine, in its crude form, appears only in the third part of the canon, the so-called Abhidharma, but not in the first and second parts, in the Sûtras, the sermons, and the Vinaya, the ethics, which together bear the name of Dharma or Law. He next points out that, according to some ancient authorities, this entire part of the canon was designated as "not pronounced by Buddha."[1] These are, at once, two important limitations. I add a third, and maintain that sayings of the Buddha occur in the first and second parts of the canon, which are in open contradiction to this metaphysical Nihilism.

Now as regards the soul, or the self, the existence of which, according to the orthodox metaphysics, is purely phenomenal, a sentence attributed to the Buddha says, "Self is the Lord of Self, who else could be the Lord?" And again, "A man who controls himself enters the untrodden land through his own self-controlled self." And this untrodden land is the Nirvâna.

Nirvâna certainly means extinction, whatever its later arbitrary interpretations may have been, and seems therefore to imply, even etymologically, a real blowing out or passing away. But Nirvâna occurs also in the Brahmanic writings, as synonymous with Moksha, Nirvritti, and other words, all designating the highest stage of spiritual liberty and bliss, but not annihilation. Nirvâna may mean the extinction of many things — of selfishness, desire, and sin, without going so far as the extinction of subjective consciousness. Further, if we consider that Buddha himself, after he had already seen Nirvâna, still remains on earth until his body falls a prey to death; that Buddha appears, in the legends, to his disciples even after his death, it seems to me that all these circumstances are hardly reconcilable with the orthodox metaphysical doctrine of Nirvâna.

What does it mean when Buddha calls reflection the path of immortality, and thoughtlessness the path of

[1] Max Müller's *Chips*, second edition, vol. i. p. 285, note.

death? Buddhaghosha, a learned man of the fifth century, here explains immortality by Nirvâna, and that this was also Buddha's thought is clearly established by a passage following immediately after: "These wise people, meditative, steady, always possessed of strong powers, attain to Nirvâna, the highest happiness." Can this be annihilation? and would such expressions have been used by the founder of this new religion, if what he called immortality had, in his own idea, been annihilation?

I could quote many more such passages did I not fear to tire you. Nirvâna occurs even in the purely moral sense of quietness and absence of passion. "When a man can bear everything without uttering a sound," says Buddha, "he has attained Nirvâna." Quiet long-suffering he calls the highest Nirvâna; he who has conquered passion and hatred is said to enter into Nirvâna.

In other passages, Nirvâna is described as the result of just knowledge. There we read: "Hunger or desire is the worst ailment, the body the greatest of all evils; where this is properly known, there is Nirvâna, the greatest happiness."

When it is said in one passage that Rest (Sânti) is the highest bliss, it is said in another that Nirvâna is the highest bliss.

Buddha says: "The sages who injure nobody, and who always control their body, they will go to the unchangeable place (Nirvâna), where, if they have gone, they will suffer no more."

Nirvâna is called the quiet place, the immortal place, even simply that which is immortal; and the expression occurs, that the wise dived into this immortal. As, according to Buddha, everything that was made, everything that was put together, passes away again, and resolves itself into its component parts, he calls in contradistinction, that which is not made, *i. e.*, the uncreated and eternal, Nirvâna. He says: "When you have understood the destruction of all that was made, you will understand that which was not made." Whence it appears that even for him a certain something exists, which is not made, which is eternal and imperishable.

On considering such sayings, to which many more might be added, one recognizes in them a conception of Nirvâna, altogether irreconcilable with the Nihilism of the third part of the Buddhist Canon. The question in such matters is not a more or less, but an *aut-aut*. If these sayings have maintained themselves, in spite of their contradiction to orthodox metaphysics, the only explanation, in my opinion, is, that they were too firmly fixed in the tradition which went back to Buddha and his disciples. What Bishop Bigandet and others represent as the popular view of the Nirvâna, in contradistinction to that of the Buddhist divines, was, if I am not mistaken, the conception of Buddha and his disciples. It represented the entrance of the soul into rest, a subduing of all wishes and desires, indifference to joy and pain, to good and evil, an absorption of the soul in itself, and a freedom from the circle of existences from birth to death, and from death to a new birth. This is still the meaning which educated people attach to it, whilst, to the minds of the larger masses,[1] Nirvâna suggests rather the idea of a Mohammedan paradise or of blissful Elysian fields.

[1] Bigandet, *The Life or Legend of Gaudama, the Buddha of the Burmese*, with Annotations. The ways to Neibban, and notice on the Phongyies, or Burmese Monks. 8vo, sewed, pp. xi., 538, and v. London, Trübner & Co.

Only in the hands of the philosophers, to whom Buddhism owes its metaphysics, the Nirvâna, through constant negations, carried to an indefinite degree, through the excluding and abstracting of all that is not Nirvâna, at last became an empty Nothing, a philosophical myth. There is no lack of such philosophical myths either in the East or in the West. What has been fabled by philosophers of a Nothing, and of the terrors of a Nothing, is as much a myth as the myth of Eos and Tithonus. There is no more a Nothing than there is an Eos or a Chaos. All these are sickly, dying, or dead words, which, like shadows and ghosts, continue to haunt language, and succeed in deceiving for a while even the healthiest understanding.

Even modern philosophy is not afraid to say that there is a Nothing. We find passages in the German mystics, such as Eckhart and Tauler, where the abyss of the Nothing is spoken of quite in a Buddhist style. If Buddha had said, like St. Paul, "that what no eye hath seen, nor ear heard, neither has it entered into the heart of man," was prepared in the Nirvâna for those who had advanced to the highest degree of spiritual perfection, such expressions would have been quite sufficient to serve as a proof to the philosophers by profession that this Nirvâna, which could not become an object of perception by the senses, nor of conception by the categories of the understanding, could be nothing more nor less than the Nothing. Could we dare with Hegel to distinguish between a Nothing (Nichts) and a Not (Nicht), we might say that the Nirvâna had through a false dialectical process become from a relative Nothing an absolute Not. This was the work of the theologians and of the orthodox philosophers. But a religion has never been founded by such teaching, and a man like Buddha, who knew mankind, must have known that he could not with such weapons overturn the tyranny of the Brahmans. Either we must bring ourselves to believe that Buddha taught his disciples two diametrically opposed doctrines on Nirvâna, say an exoteric and esoteric one, or we must allow *that* view of Nirvâna to have been the original view of the founder of this marvelous religion, which corresponds best with the simple, clear, and practical character of Buddha.

I have now said all that can be said in vindication of Buddha within the brief time allowed to these discourses. But I should be sorry if you carried away the impression that Buddhism contained nothing but empty, useless speculations; permit me, therefore, to read to you, in conclusion, a short Buddhist parable, which will show you Buddhism in a more human form. It is borrowed from a work which will soon appear, and which contains the translation of the parables used by the Buddhists to obtain acceptance for their doctrines amongst the people. I shall only omit some technical expressions and minor details which are of no importance:[1] —

"Some time after this, Kisâgotamî gave birth to a son. When the boy was able to walk by himself, he died. The young girl, in her love for it, carried the dead child clasped to her bosom, and went about from house to house, asking if any one would give her some medicine for it. When the neighbors saw this, they said, 'Is the young girl mad that she carries about on her breast the dead body of her son!' But a wise man thinking to himself, 'Alas! this Kisâgotamî does not understand the law of

---

[1] *Buddhaghosha's Parables.* Translated from the Burmese by Captain H. T. Rogers, R. E. With an Introduction containing "Buddha's Dhammapada, or the Path of Virtue." Translated from Pâli by Professor F. Max Müller. London, Trübner & Co.

death, I must comfort her,' said to her, 'My good girl, I cannot myself give medicine for it, but I know of a doctor who can attend to it.' The young girl said, 'If so, tell me who it is.' The wise man continued, 'Buddha can give medicine, you must go to him.'

"Kisâgotamî went to Buddha, and doing homage to him, said, 'Lord and master, do you know any medicine that will be good for my boy?' Buddha replied, 'I know of some.' She asked, 'What medicine do you require?' He said, 'I want a handful of mustard seed.' The girl promised to procure it for him, but Buddha continued, 'I require some mustard seed taken from a house where no son, husband, parent, or slave has died.' The girl said, 'Very good,' and went to ask for some at the different houses, carrying the dead body of her son astride on her hip. The people said, 'Here is some mustard seed, take it.' Then she asked, 'In my friend's house has there died a son, a husband, a parent, or a slave?' They replied, 'Lady, what is this that you say! *The living are few, but the dead are many.*' Then she went to other houses, but one said, 'I have lost a son;' another, 'I have lost my parents;' another, 'I have lost my slave.' At last, not being able to find a single house where no one had died, from which to procure the mustard seed, she began to think, 'This is a heavy task that I am engaged in. I am not the only one whose son is dead. In the whole of the Sâvatthi country, everywhere children are dying, parents are dying.' Thinking thus, she was seized by fear, and putting away her affection for her child, she summoned up resolution, and left the dead body in a forest; then she went to Buddha and paid him homage. He said to her, 'Have you procured the handful of mustard seed?' 'I have not,' she replied; 'the people of the village told me, "*The living are few, but the dead are many.*"' Buddha said to her, 'You thought that you alone had lost a son; the law of death is that among all living creatures there is no permanence.' When Buddha had finished preaching the law, Kisâgotamî was established in the reward of the noviciate; and all the assembly who heard the law were established in the same reward.

"Some time afterwards, when Kisâgotamî was one day engaged in the performance of her religious duties, she observed the lights (in the houses) now shining, now extinguished, and began to reflect, 'My state is like these lamps.' Buddha, who was then in the Gandhakuṭi building, sent his sacred appearance to her, which said to her, just as if he himself was preaching, 'All living beings resemble the flame of these lamps, one moment lighted, the next extinguished; those only who have arrived at Nirvâna are at rest.' Kisâgotami, on hearing this, reached the stage of a saint possessed of intuitive knowledge."

Gentlemen, this is a specimen of the true Buddhism; this is the language, intelligible to the poor and the suffering, which has endeared Buddhism to the hearts of millions, — not the silly, metaphysical phantasmagorias of worlds of gods and worlds of Brahma, or final dissolution of the soul in Nirvâna, — no, the beautiful, the tender, the humanly true, which, like pure gold, lies buried in all religions, even in the sand of the Buddhist Canon.

BUDDHA'S DHAMMAPADA,

OR

"PATH OF VIRTUE."

Translated from Pâli

By F. MAX MÜLLER, M. A.,

PROFESSOR OF COMPARATIVE PHILOLOGY AT OXFORD, FOREIGN
MEMBER OF THE FRENCH INSTITUTE, ETC.

BUDDHA'S DHAMMAPADA,

OR

"PATH OF VIRTUE."

[The accompanying essay upon the age of Buddhaghosha's Parables and of the Dhammapada by Prof. Max Müller originally appeared as a preface to a translation from the Parables themselves by Capt. T. Rogers, R. E. A few introductory paragraphs relating particularly to Capt. Rogers and his share of the volume are omitted from this prefatory essay.—THE PUBLISHERS.]

THE Dhammapada forms part of the Buddhistic Canon, and consists of four hundred and twenty-four verses,[1] which are believed to contain the utter-

---

[1] That there should be some differences in the exact number of these gâthâs, or verses, is but natural. In a short index at the end of the work, the number of chapters is given as twenty-six. This agrees with our text. The sum total, too, of the verses as there given, namely 423, agrees with the number of verses which Buddhaghosha had before him, when writing his commentary, at the beginning of the fifth century of our era. It is only when the number of verses in each chapter is given that some slight differences occur. Cap. v. is said to contain 17 instead of 16 verses; cap. xii. 12 instead of 10; cap xiv. 16 instead of 18; cap. xx. 16 instead of 17; cap. xxiv. 22 instead of 26; cap. xxvi. 40 instead of 41, which would give altogether five verses less than we actually possess. The cause of this difference may be either in the wording of the index itself (and we actually find it in a various reading, malavagge ka vîsati, instead of malavagg' ekavîsati, see Fausböll, p. 435); or in the occasional counting of two verses as one, or of one as two. Thus in cap. v. we get 16 instead of 17 verses, if we take each verse to consist of two lines only, and not, as in vv. 74 and 75, of three. Under all circumstances the difference is trifling, and we may be satisfied that we possess in our MSS. the same text which Buddhaghosha knew in the fifth century of our era.

ances of Buddha himself. It is in explaining these verses that Buddhaghosha gives for each verse a parable, which is to illustrate the meaning of the verse, and is believed to have been uttered by Buddha, in his intercourse with his disciples, or in preaching to the multitudes that came to hear him. In translating these verses, I have followed the edition of the Páli text, published in 1855 by Dr. Fausböll, and I have derived great advantage from his Latin translation, his notes, and his copious extracts from Buddhaghosha's commentary. I have also consulted translations, either of the whole of the Dhammapada, or of portions of it, by Weber, Gogerly,[1] Upham, Burnouf, and others. Though it will be seen that in many places my translation differs from those of my predecessors, I can only claim for myself the name of a very humble gleaner in the field of Páli literature. The greatest credit is due to Dr. Fausböll, whose *editio princeps* of the Dhammapada will mark forever an important epoch in the history of Páli scholarship; and though later critics have been able to point out some mistakes, both in his text and in his translation, the value of their labors is not to be compared with that of the work accomplished single-handed by that eminent Danish scholar.

ON THE AGE OF THE PARABLES AND OF THE DHAMMAPADA.

The age of Buddhaghosha can be fixed with greater accuracy than most dates in the literary history of India, for not only his name, but the circumstances of his life and his literary activity are described in the

---

[1] "Several of the chapters have been translated by Mr. Gogerly, and have appeared in *The Friend*, vol. iv. 1840." Spence Hardy, *Eastern Monachism*, p. 169.

Mahâvansa, the history of Ceylon, by what may be called almost a contemporary witness. The Mahâvansa, lit. the genealogy of the great,[1] or the great genealogy, is up to the reign of Dhâtusena, the work of Mahânâma. It was founded on the Dîpavansa, also called Mahâvansa, a more ancient history of the island of Ceylon, which ended with the reign of Mahâsena, who died 302 A. D. MSS. of the Dîpavansa are said to exist, and there is a hope of its being published. Mahânâma, who lived during the reign of King Dhâtusena, 459–477, wrote the whole history of the island over again, and carried it on to his own time. He also wrote a commentary on this work, but that commentary extends only as far as the forty-eighth verse of the thirty-seventh chapter, *i. e.*, as far as the reign of Mahâsena, who died in 502 A. D.[2] As it breaks off exactly where the older history, the Dîpavansa, is said to have ended, it seems most likely that Mahânâma embodied in it the results of his own researches into the ancient history of Ceylon, while for his continuation of the work, from the death of Mahâsena to his own time, no such commentary was wanted. It is difficult to determine whether the thirty-eighth as well as the thirty-seventh chapter came from the pen of Mahânâma, for the Mahâvansa was afterwards continued by different writers to the middle of the last century; but, taking into account all the circumstances of the case, it is most probable

---

[1] See Mahânâma's own explanations given in the Tîkâ; *Mahavansa*, Introduction, p. xxxi.

[2] After the forty-eighth verse, the text, as published by Turnour, puts "Mahâvanso nitthito," the Mahâvansa is finished; and after a new invocation of Buddha, the history is continued with the forty-ninth verse. The title Mahâvansa, as here employed, seems to refer to the Dîpavansa.

that Mahânâma carried on the history to his own time, to the death of Dhâtusena or Dâsen Kellîya, who died in 477.[1] This Dhâtusena was the nephew of the historian Mahânâma, and owed the throne to the protection of his uncle. Dhâtusena was in fact the restorer of a national dynasty, and after having defeated the foreign usurpers (the Damilo dynasty) "he restored the religion which had been set aside by the foreigners."[2] Among his many pious acts, it is particularly mentioned that he gave a thousand, and ordered the Dîpavansa to be promulgated.[3]

As Mahânâma was the uncle of Dhâtusena, who reigned from 459–477, he may be considered a trustworthy witness with regard to facts that occurred between 410 and 432. Now the literary activity of Buddhaghosha in Ceylon falls in that period, and this is what Mahânâma relates of him:[4]—

"A Brâhman youth, born in the neighborhood of the terrace of the great Bo-tree (in Mâgadha), accomplished in the 'vijjá' (knowledge) and 'sippa' (art), who had achieved the knowledge of the three Vedas, and possessed great aptitude in attaining acquirements; indefatigable as a schismatic disputant, and himself a schismatic wanderer over Gambudipa, established himself, in the character of a disputant, in a certain vihâra, and was in the habit of rehearsing, by night and by day with clasped hands, a discourse which he had learned, perfect in all its component parts, and sustained throughout in the same lofty strain.

---

[1] *Mahâvansa*, Introduction, p. xxxi.
[2] Ibid. p. 256.
[3] Ibid. p. 257, "And that he might also promulgate the contents of the Dîpavansa, distributing a thousand pieces, he caused it to be read aloud thoroughly." The text has, "datvâ sahassam dîpetum Dîpavansam samâdisi," having given a thousand, he ordered the Dîpavansa to be rendered illustrious, or to be copied. (See Westergaard, *Ueber den ältesten Zeitraum der Indischen Geschichte*, Breslau, 1862, p. 33; and *Mahâvansa*, Introduction, p. xxxii. l. 2.)
[4] *Mahâvansa*, p. 250.

A certain mahâ'hera, Revata, becoming acquainted with him there, and (saying to himself), 'This individual is a person of profound knowledge; it will be worthy (of me) to convert him;' inquired, 'Who is this who is braying like an ass?' The Brâhman replied to him, 'Thou canst define, then, the meaning conveyed in the bray of asses.' On the Thera rejoining, 'I can define it;' he (the Brâhman) exhibited the extent of the knowledge he possessed. The Thera criticised each of his propositions, and pointed out in what respect they were fallacious. He who had been thus refuted, said, 'Well, then, descend to thy own creed;' and he propounded to him a passage from the Abhidhamma (of the Pitakattaya). He (the Brâhman) could not divine the signification of that passage, and inquired, 'Whose manta is this?' — 'It is Buddha's manta.' On his exclaiming, 'Impart it to me;' the Thera replied, 'Enter the sacerdotal order.' He who was desirous of acquiring the knowledge of the Pitakattaya, subsequently coming to this conviction, 'This is the sole road' (to salvation), became a convert to that faith. As he was as profound in his eloquence (ghosa) as Buddha himself, they conferred on him the appellation of Buddhaghosa (the voice of Buddha); and throughout the world he became as renowned as Buddha. Having there (in *G*ambudipa) composed an original work called 'Nânodaya' (Rise of Knowledge), he, at the same time, wrote the chapter called 'Atthasâlini, on the Dhammasanganî' (one of the Commentaries on the Abhidhamma).

"Revata Thera then observing that he was desirous of undertaking the compilation of a general commentary on the Pitakattaya, thus addressed him: 'The text alone of the Pitakattaya has been preserved in this land, the Atthakathâ are not extant here, nor is there any version to be found of the schisms (vâda) complete. The Singhalese Atthakathâ are genuine. They were composed in the Singhalese language by the inspired and profoundly wise Mahinda, who had previously consulted the discourses of Buddha, authenticated at the thera-convocations, and the dissertations and arguments of Sâriputta and others, and they are extant among the Singhalese. Preparing for this, and studying the same, translate them according to the rules of the grammar of the Mâgadhas. It will be an act conducive to the welfare of the whole world.'

"Having been thus advised, this eminently wise personage

rejoicing therein, departed from thence, and visited this island in the reign of this monarch (*i. e.* Mahânâma). On reaching the Mahâvihâra (at Anurâdhapura), he entered the Mahâpadhâna hall, the most splendid of the apartments in the vihâra, and listened to the Singhalese Atthakathâ, and the Theravâda, from the beginning to the end, propounded by the thera Sanghapâla ; and became thoroughly convinced that they conveyed the true meaning of the doctrines of the Lord of Dhamma. Thereupon paying reverential respect to the priesthood, he thus petitioned : ' I am desirous of translating the Atthakathâ; give me access to all your books.' The priesthood, for the purpose of testing his qualifications, gave only two gâthâs, saying, ' Hence prove thy qualification ; having satisfied ourselves on this point, we will then let thee have all our books.' From these (taking these gâthâ for his text), and consulting the Pitakattaya, together with the Atthakathâ, and condensing them into an abridged form, he composed the work called ' The Visuddhimagga.' Thereupon, having assembled the priesthood, who had acquired a thorough knowledge of the doctrines of Buddha, at the bo-tree, he commenced to read out the work he had composed. The devatâs, in order that they might make his (Buddhaghosa's) gifts of wisdom celebrated among men, rendered that book invisible. He, however, for a second and third time recomposed it. When he was in the act of producing his book for the third time, for the purpose of propounding it, the devatâs restored the other two copies also. The assembled priests then read out the three books simultaneously. In those three versions, neither in a signification nor in a single misplacement by transposition, nay even in the thera-controversies, and in the text (of the Pitakattaya) was there, in the measure of a verse or in the letter of a word, the slightest variation. Thereupon, the priesthood rejoicing, again and again fervently shouted forth, saying, ' Most assuredly this is Metteya (Buddha) himself,' and made over to him the books in which the Pitakattaya were recorded, together with the Atthakathâ. Taking up his residence in the secluded Ganthâkara vihâra, at Anurâdhapura, he translated, according to the grammatical rules of the Mâgadhas, which is the root of all languages, the whole of the Singhalese Atthakathâ (into Pâli). This proved an achievement of the utmost consequence to all languages spoken by the human race.

" All the theras and âchâriyas held this compilation in the

same estimation as the text (of the Pitakattaya). Thereafter, the objects of his mission having been fulfilled, he returned to *G*ambudipa, to worship at the bo-tree (at Uruvelâya, or Uruvilvâ, in Mâgadha)."

Here we have a simple account of Buddhaghosha[1] and his literary labors written by a man, himself a priest, and who may well have known Buddhaghosha during his stay in Ceylon. It is true that the statement of his writing the same book three times over without a single various reading, partakes a little of the miraculous; but we find similar legends mixed up with accounts of translations of other sacred books, and we cannot contend that writers who believed in such legends are therefore unworthy to be believed as historical witnesses.

The next question which has to be answered is this, Did Buddhaghosha's Parables, and the whole of the commentary in which they are contained, form part of the Arthakathâ which he translated from Singhalese into Pâli. The answer to this question depends on whether the Dhammapada formed part of the Pi-

[1] The Burmese entertain the highest respect for Buddhaghosha. Bishop Bigandet, in his *Life or Legend of Gaudama* (Rangoon, 1866), writes: "It is perhaps as well to mention here an epoch which has been, at all times, famous in the history of Buddhism in Burma. I allude to the voyage which a Religious of Thaton, named Budhagosa, made to Ceylon, in the year of religion 943=400 A. C. The object of this voyage was to procure a copy of the scriptures. He succeeded in his undertaking. He made use of the Burmese, or rather Talaing characters, in transcribing the manuscripts, which were written with the characters of Magatha. The Burmans lay much stress upon that voyage, and always carefully note down the year it took place. In fact, it is to Budhagosa that the people living on the shores of the Gulf of Martaban owe the possession of the Budhist scriptures. From Thaton, the collection made by Budhagosa was transferred to Pagan, six hundred and fifty years after it had been imported from Ceylon."

takattaya or not. If the verses of the Dhammapada were contained in the canon, then they were also explained in the Singhalese Arthakathâ, and consequently translated from it into Pâli by Buddhaghosha. Now it is true that the exact place of the Dhammapada in the Buddhistic Canon has not yet been pointed out; but if we refer to Appendix iii., printed in Turnour's edition of the Mahâvansa, we there find in the third part of the canon, the Sûtra-pitaka, under No. 5, the Kshudraka-nikâya, containing fifteen subdivisions, the second of which is the Dhammapada.

We should, therefore, be perfectly justified in treating the parables contained in Buddhaghosha's Pâli translation of the Arthakathâ, *i. e.* the commentary on the Dhammapada, as part of a much more ancient work, namely, the work of Mahinda, and it is only in deference to an over-cautious criticism that I have claimed no earlier date than that of Buddhaghosha for these curious relics of the fable-literature of India. I have myself on a former occasion[1] pointed out all the objections that can be raised against the authority of Buddhaghosha and Mahinda; but I do not think that scholars calling these parables the parables of Mahinda, if not of Buddha himself, and referring their date to the third century B. C., would expose themselves at present to any formidable criticism.

If we read the pages of the Mahâvansa without prejudice, and make allowance for the exaggerations and superstitions of Oriental writers, we see clearly that the literary work of Buddhaghosha presupposes the existence, in some shape or other, not only of the canonical books, but also of their Singhalese commentary. The Buddhistic Canon had been settled in sev-

---

[1] *Chips from a German Workshop*, 2d ed. vol. i. p. 197.

eral councils, whether two or three, we need not here inquire.[1] It had received its final form at the council held under Asoka in the year 246 B. C. We are further told in the Mahâvansa that Mahinda, the son of Asoka, who had become a priest, learnt the whole of the Buddhist Canon in three years;[2] and that at the end of the third council he was dispatched to Ceylon, in order to establish there the religion of Buddha.[3] The king of Ceylon, Devânâmpriya Tishya, was converted, and Buddhism soon became the dominant religion of the island. Next follows a statement which will naturally stagger those who are not acquainted with the power of memory if under strict discipline for literary purposes, but which exceeds by no means the limits of what is possible in times when the whole sacred literature of a people is preserved, and lives by oral tradition only. The Pitakattaya, as well as the Arthakathâ, having been collected and settled at the third council in 246 B. C., were brought to Ceylon by Mahinda, who promulgated them orally;[4] the Pitakattaya in Pâli, and the Arthakathâ in Singhalese,[5] together with additional Arthakathâ of his own. It

[1] The question of these councils and of their bearing on Indian chronology has been discussed by me in my *History of Ancient Sanskrit Literature*, p. 262 seq., 2d ed.
[2] *Mahâvansa*, p. 37.   [3] Ibid. p. 71.   [4] Cf. Bigandet, l. c. p. 387.
[5] Singhalese, being the language of the island, would naturally be adopted by Mahinda and his fellow-missionaries for communication with the natives. If he abstained from translating the canon also into Singhalese, this may have been on account of its more sacred character. At a later time, however, the canon, too, was translated into Singhalese, and, as late as the time of Buddhadâsa, who died 368 A. D., we read of a priest, profoundly versed in the doctrines, who translated the Sûtras, one of the three divisions of the Pitakattaya, into the Sihala language. *Mahâv.* p. 247. A note is added, stating that several portions of the other two divisions also of the Pitakattaya have been translated into the Singhalese language, and that these alone are

does not follow that Mahinda knew the whole of that enormous literature by heart, for, as he was supported by a number of priests, they may well have divided the different sections among them. The same applies to their disciples. But that to the Hindu mind there was nothing exceptional or incredible in such a statement, we see clearly from what is said by Mahânâma at a later period of his history. When he comes to the reign of Va*tt*agâmani,[1] 88–76 B. C., he states: "The profoundly wise priests had heretofore orally perpetuated the Pâli Pitakattaya and its Arthakathâ (commentaries). At this period these priests, foreseeing the perdition of the people (from the perversions of the true doctrines) assembled; and in order that the religion might endure for ages, recorded the same in books."[2]

Later than this date, even those who doubt the powers of oral tradition have no right to place the final constitution of the Buddhistic Canon and its commentaries in Ceylon, nor is there any reason to doubt that such as these texts existed in Ceylon in the first century B. C., they existed in the fifth century after Christ,

---

consulted by the priests who are unacquainted with Pâli. On the other hand, it is stated that the Singhalese text of the Arthakathâ exists no longer (see Spence Hardy, *Legends*, p. xxv., and p. 69). He states that the text and commentary of the Buddhist Canon are believed to contain 29,368,000 letters. Ibid. p. 66.

[1] See Bigandet, l. c. p. 388.

[2] See also Spence Hardy, *Legends*, p. 192. "After the Nirvâna of Buddha, for the space of four hundred and fifty years, the text and commentaries. and all the works of the Tathâgata, were preserved and transmitted by wise priests, orally, mukha-pâ*th*ena. But having seen the evils attendant upon this mode of transmission, five hundred and fifty arhats, of great authority, in the cave called Aloka (Alu) in the province of Malaya, in Lankâ, under the guardianship of the chief of that province, caused the (sacred) books to be written." — *Extract from the "Sâra-sangraha."*

when the commentaries were translated into Pâli by Buddhaghosha, and that afterwards they remained unchanged in the MSS. preserved by the learned priests of that island. It is easy to shrug one's shoulders, and shake one's head, and to disbelieve everything that can be disbelieved. Of course we cannot bring witnesses back from the grave, still less from the Nirvâna, into which, we trust, many of these ancient worthies have entered. But if we are asked to believe that all this was invented in order to give to the Buddhistic Canon a fictitious air of antiquity, the achievement would, indeed, be one of consummate skill. When Asoka first met Nigrodha, who was to convert him to the new faith, we read,[1] that having refreshed the saint with food and beverage which had been prepared for himself, he interrogated the sâmanera on the doctrines propounded by Buddha. It is then said that the sâmanera explained to him the Apramâda-varga. Now this Apramâda-varga is the title of the second chapter of the Dhammapada. Its mention here need not prove that the Dhammapada existed previous to the Council of Asoka, 246 B. C., but only that Mahânâma believed that it existed before that time. But if we are to suppose that all this was put in on purpose, would it not be too deep-laid a scheme for the compiler of the Mahâvansa?[2]

And for what object could all this cunning have been employed? The Buddhists would have believed the most miraculous accounts that might be given of the origin and perpetuation of their sacred writings; why then tell the story so plainly, so baldly, so simply

---

[1] *Mahâvansa*, p. 25.

[2] In the account given by Bishop Bigandet (p. 377) of the first interview between Asoka and Nigrodha, the lines repeated by the priest to the king are likewise taken from the Apramâda-varga.

as a matter of fact? I have the greatest respect for really critical skepticism, but a skepticism without any arguments to support it is too cheap a virtue to deserve much consideration. Till we hear some reasons to the contrary, I believe we may safely say that we possess Buddhaghosha's translation of the Arthakathâ as it existed in the fifth century of our era; that the original was first reduced to writing in Ceylon in the first century before our era, having previously existed in the language of Magadha; and that our verses of the Dhammapada are the same which were recited to Asoka, and embodied in the canon of the third council, 246 B.C. This is enough for our purposes: the chronology previous to Asoka, or at least previous to his grandfather, *K*andragupta, the ally of Seleucus, belongs to a different class of researches.

As, however, the antiquity and authenticity of the Buddhist literature have of late been called in question in a most summary manner, it may not seem superfluous to show, by one small fact at least, that the fables and parables of Buddhaghosha must have existed *in the very wording in which we possess them*, in the beginning, at least, of the sixth century of our era. It was at that time that Khosru Anushirván (531–579) ordered a collection of fables[1] to be translated from Sanskrit into the language of Persia, which translation became in turn the source of the Arabic and the other numerous translations of that ancient collection of apologues. These Sanskrit fables as collected in the Pañ*k*atantra, have been proved by Prof. Benfey to have been borrowed from Buddhistic sources; and I believe we may go even a step further and maintain, that not only the general outlines of these fables, but in some cases the very words, were taken over from Pâli into Sanskrit.

[1] See Benfey, *Pantschatantra*, vol. i. p. 6.

We read in the Pankatantra, ii. 10, the following verse: —

> Gâlam âdâya gakkhanti sahasâ [1] pakshino 'py amî,
> Yâvak ka vivadishyante patishyanti na samsayah.

"Even these birds fly away quickly taking the net; and when they shall quarrel, they will fall, no doubt."

This verse recapitulates the story of the birds which are caught in a net, but escape the fowler by agreeing to fly up together at the same moment. The same story is told in the Hitopadesa, i. 36 (32): —

> Samhatâs tu haranty ete mama gâlam vihamgamâh,
> Yadâ tu nipatishyanti vasam eshyanti me tadâ.

"Combined indeed do these birds take away my net; but when they fall down, they will then fall into my power."

The first thing that should be pointed out is, that of these two versions of the same idea, neither is borrowed from the other, neither that of the Hitopadesa from the Pañkantra, nor *vice versâ*.[2] They presuppose a common source from which they are derived, thus sharing together certain terms in common, and following an independent course in other respects. This common source is a Pâli verse which occurs in

---

[1] If we read "samhatâh" instead of "sahasâ," we have to translate, "Holding together even these birds fly away, taking the net."

[2] A third version is found in the Mahâbhârata, *Udoyaga-parva*, v. 2461, where a similar story is told of two birds being caught and escaping from the fowler by agreeing to fly up together. Here we read: —

> Pâsam ekam ubhâv etam sahitau harato mama,
> Yatra vai vivadishyete tatra me vasam eshyatah.

"These two united carry off this one net of mine; when they shall quarrel, then they will fall into my power."

the Vattaka-*g*âtaka, and is quoted by Buddhaghosha in his commentary on the Sûtra-nipâta.[1]

>Sammodamânâ ga*kkh*anti *g*âlam âdâya pakkhino,
>Yadâ te vivadissanti tadâ ehinti me vasa*m*.

"The birds fly away, taking the net while they are happy together; when they shall quarrel, then they will come into my power."

If we mark these three verses by the letters P., H., and V., we see that P. takes from V. the words "*g*âlam âdâya ga*kkh*anti pakshina*h*" and "vivadishyante," while H. takes from V. the words "vasam eshyanti me tadâ." For the rest, H. and P. follow each their own way in transforming the Pâli verse, as best they can, into a Sanskrit verse, and H. with more success than P. The words "apy amî" in P. are mere expletives, "patishyanti" is a poor rendering, and "na sam*s*aya*h*" again is added only in order to fill the verse. Without calling H. and P. together a faithful copy of V., I think we may safely say that it would be impossible to explain both the points on which H. and P. differ and those on which they agree, without admitting that both had before them the Pâli verse in the very wording in which we find it in Buddhaghosha's commentary, and which, according to Buddhaghosha, was taken from one of the *G*âtakas, a portion of the Buddhistic Canon. And this would prove, though one could hardly have thought that, after the labors of Burnouf and Lassen and Julien,[2] such proof was still

[1] This extract from the commentary was published by Dr. Fausböll in the *Indische Studien*, vol. v. p. 412, and the similarity was pointed out between the verse of Buddhaghosha and the corresponding verses in the Hitopadesa and Pañkatantra. Further comparisons may be seen in Benfey, *Pañkatantra*, vol. i. p. 305; vol. ii. pp. 450, 540. See also *Les Avadânas traduits par Stanislas Julien*, vol. i. p. 155.

[2] On Buddhist books carried to China and translated there previous to the beginning of our era, see M. M.'s *Chips from a German Workshop*, 2d ed. vol. i. p. 258, *seq.*

needed, that the Buddhist Canon and its commentary existed in the very wording in which we now possess them, previous at least to 500 after Christ.

ON THE IMPORTANCE OF THE DHAMMAPADA.

If we may consider the date of the Dhammapada firmly established, and treat its verses, if not as the utterances of Buddha, at least as what were believed by the members of the council under Asoka, in 246 B. C., to have been the utterances of the founder of their religion, its importance for a critical study of the history of Buddhism must be very considerable, for we can hardly ever expect to get nearer to Buddha himself and to his personal teaching. I shall try to illustrate this by one or two examples.

I pointed out on a former occasion [1] that if we derive our ideas of Nirvâna from the Abhidharma, i. e., the metaphysical portion of the Buddhistic Canon, we cannot escape the conclusion that that it meant perfect annihilation. Nothing has been brought forward to invalidate Burnouf's statements on this subject, much has since been added, particularly by M. Barthélemy St. Hilaire, to strengthen and support them, and the latest writer on Buddhism, Bishop Bigandet, the Vicar Apostolic of Ava and Pegu, in his "Life and Legend of Gaudama, the Buddha of the Burmese," arrives at exactly the same conclusion. No one could suspect the bishop of any prejudice against Buddhism, for he is most candid in his praises of whatever is praiseworthy in that ancient system of religion. Thus he says [2] "The Christian system and the Buddhistic one, though differing from each other in their respect-

---

[1] On the meaning of Nirvâna, in *Chips from a German Workshop*, 2d ed. vol. i. p. 280.   [2] Page 494.

ive objects and ends as much as truth from error, have, it must be confessed, many striking features of an astonishing resemblance. There are many moral precepts equally commanded and enforced in common by both creeds. It will not be considered rash to assert that most of the moral truths prescribed by the gospel are to be met with in the Buddhistic scriptures." And again,[1] "In reading the particulars of the life of the last Budha Gautama, it is impossible not to feel reminded of many circumstances relating to our Saviour's life, such as it has been sketched by the Evangelists." Yet, in spite of all these excellences, Bishop Bigandet, too, sums up dead against Budhism, as a religion culminating in atheism and nihilism. "It may be said in favor of Buddhism," he writes,[2] "that no philosophico-religious system has ever upheld, to an equal degree, the notions of a savior and deliverer, and the necessity of his mission for procuring the salvation, in a Buddhistic sense, of man. The *rôle* of Buddha, from beginning to end, is that of a deliverer, who preaches a law designed to procure to man the deliverance from all the miseries he is laboring under. By an inexplicable and deplorable eccentricity, the pretended savior, after having taught man the way to deliver himself from the tyranny of his passions, leads him, after all, into the bottomless gulf of 'total annihilation.'"

That Buddha was an atheist, at least in one sense of the word, cannot be denied, but whether he believed in a total annihilation of the soul as the highest goal of religion, is a different question. The gods whom he found worshipped by the multitude, were the gods of the Vedas and the Brâhmanas, such as Indra, Agni,

[1] Page 495. [2] Page viii.

and Yama, and in the divinity of such deities, Buddha certainly did not believe. He never argues against their existence; on the contrary, he treats the old gods as superhuman beings, and promises his followers who have not yet reached the highest knowledge, but have acquired merit by a virtuous life, that after death they shall be born again in the world of the gods, and enjoy divine bliss in company with these deities. Similarly he threatens the wicked that after death they shall meet with their punishment in the subterranean abodes and hells, where Asuras, Sarpas, Pretas, and other spirits dwell. The belief in these beings was so firmly rooted in the popular belief and language that even the founder of a new religion could not have dared to reason them away, and there was so little in the doctrine of Buddha that appealed to the senses or lent itself to artistic representation, whether in painting or sculpture, that nothing remained to Buddhist artists but to fall back for their own purposes on the old mythology, or at least on the popular superstition, the fairy and snake tales of the people.[1]

[1] This may be seen from the curious ornamentations of Buddhist temples, some of which were lately published by Mr. Fergusson. Those of the Sanchi tope are taken from drawings executed for the late East India Company by Lieutenant (now Lieutenant-colonel) Maisey, and from photographs by Lieutenant Waterhouse; those of the Amravatî tope are photographed from the sculptured slabs sent home by Colonel Mackenzie, formerly exhibited in the Museum of the East India Company, and from another valuable collection sent home by Sir Walter Elliot. Architectural evidence is supposed to fix the date of the Sanchi topes from about 250-100 B. C.; that of the gateways in the first century A. D.; while the date of the Amravatî buildings is referred to the fourth century A. D. No one would venture to doubt Mr. Fergusson's authority within the sphere of architectural chronology, but we want something more than mere affirmation when he says (p. 56), "that the earliest of the (Buddhist) scriptures we have were not reduced to writing in their present form before the fifth century after Christ."

The gods, in general, are frequently mentioned in the Dhammapada: —

V. 177. The uncharitable do not go to the world of the gods.

V. 224. Speak the truth, do not yield to anger; give, if thou art asked, from the little thou hast; by those steps thou wilt go near the gods.

V. 417. He who, after leaving all bondage to men, has risen above all bondage to the gods, him I call indeed a Brâhmana.

In vv. 44 and 45 three worlds are mentioned, the earth, the world of Yama (the lord of the departed), and the world of the gods; and in v. 126 we find hell (niraya), earth, heaven (svarga), and Nirvâna.

In v. 56 it is said that the odor of excellent people rises up to the gods; in vv. 94 and 181, that the gods envy him whose senses have been subdued; in v. 366, that they praise a Bhikshu who is contented, pure, and not slothful (cf. v. 230); in v. 224, that good people go near the gods; in v. 236, that a man who is free from guilt will enter into the heavenly world of the elect (the ariya); while in v. 187 we read of heavenly pleasures that fail to satisfy the disciples of Buddha.

Individual deities, too, are mentioned. Of Indra, who is called Maghavan, it is said in v. 30, that by perseverance he rose to the lordship of the gods.[1] In vv. 107 and 392 the worship of Agni, or fire, is spoken of as established among the Brahmans. Yama, as the lord of the departed, occurs in vv. 44, 237, and he seems to be the same as Ma*kk*urâga, the king of death, mentioned in vv. 45, 170. The men or messengers

---

[1] There is a curious story of Buddha dividing his honors with Sakka (*S*akra) or Indra on p. 162 of the *Parables*.

of Yama are spoken of in v. 235; death itself is represented as Antaka, vv. 48, 288, or as Ma*kk*u; in v. 46 the king of death (ma*kk*urâga) is mentioned together with Mâra; in v. 48 he seems to be identified with Mâra, the tempter (v. 48, note).

This Mâra, the tempter, the great antagonist of Buddha, as well as of his followers, is a very important personage in the Buddhist scriptures. He is in many places the representative of evil, the evil spirit, or, in Christian terminology, the devil, conquered by Buddha, but not destroyed by him. In the Dhammapada his character is less mythological than in other Buddhist writings. His retinue is, however, mentioned (v. 175), and his flower-pointed arrow (v. 46) reminds one of the Hindu god of love. We read that Mâra will overcome the careless, but not the faithful (vv. 7, 8, 57); that men try to escape from his dominion (v. 34), and his snares (vv. 37, 276, 350); that he should be attacked with the weapon of knowledge (v. 40); that the wise, who have conquered him, are led out of this world (v. 175). In vv. 104 and 105 we find a curious climax, if it is intended as such, from a god to a Gandharva, thence to Mâra, and finally to Brahman, all of whom are represented as powerless against a man who has conquered himself. In v. 230, too, Brahman is mentioned, and, as it would seem, as a being superior to the gods.

But although these gods and demons were recognized in the religion of Buddha, and had palaces, gardens, and courts assigned to them, hardly inferior to those which they possessed under the old *régime*, they were deprived of all their sovereign rights. Although, according to the Buddhists, the worlds of the gods last for millions of years, they must perish at the end of every kalpa with the gods and with the spirits who, in the circle of births, have raised themselves to the world of the gods. Indeed, the reorganization of the spirit-world in the hands of Buddha goes further still. Already before Buddha, the Brahmans had left the low stand-point of mythological polytheism, and had risen to the conception of the Brahman, as the absolute divine, or super-divine being. To this Brahman also, who, in the Dhammapada, already appears as superior to the gods, a place is assigned in the Buddhist demonology. Over and above the world of the gods with its six paradises, the sixteen Brahma-worlds are erected — worlds, not to be attained through virtue, and piety only, but through inner contemplation, through knowledge and enlightenment.

The dwellers in these Brahma-worlds are more than gods; they are spiritual beings, without body, without weight, without desires. Nay, even this is not sufficient, and as the Brahmans had imagined a higher Brahman, without form and without suffering (tato yad uttarataram tad arûpam anâmayam, Svet. Up. 3, 10), the Buddhists, too, in their ideal dreams, imagined four other worlds towering high above the worlds of Brahman, which they call Arûpa, the worlds of the Formless. All these worlds are open to man, after he has divested himself of all that is human, and numberless beings are constantly ascending and descending in the circle of time, according to the works they have performed, and according to the truths they have discovered. But in all these worlds the law of change prevails, in none is there exemption from birth, age, and death. The world of the gods will perish like that of men; the world of Brahman will vanish like that of the gods; nay, even the world of the Formless will not last forever; but the Buddha, the enlightened and truly free, stands higher, and will not be affected or disturbed by the collapse of the universe, Si fractus illabatur orbis, impavidum ferient ruinæ.

Here, however, we meet with a vein of irony, which one would hardly have expected in Buddha. Gods and devils he has located, to all mythological and philosophical acquisitions of the past he had done justice as far as possible. Even fabulous beings, such as Nâgas, Gandharvas, and Garudas, had escaped the process of dissolution and sublimization which was to reach them later at the hands of comparative mythologists. There is only one idea, the idea of a personal Creator, in regard to which Buddha seems merciless. It is not only denied, but even its origin, like that of an ancient myth, is carefully explained by him with the minutest detail. The Rev. D. J. Gogerly, in his numerous articles published in the local journals of Ceylon, has collected and translated the most important passages from the Buddhist Canon bearing on this subject. The Rev. Spence Hardy,[1] too, another distinguished missionary in Ceylon, has several times touched on this point — a point, no doubt, of great practical importance to Christian missionaries. They dwell on such passages as when Buddha said to Upâsaka, an ascetic, who inquired who was his teacher and whose doctrine he embraced, "I have no teacher; there is no one who resembles me. In the world of the gods I have no equal. I am the most noble in the world, being the irrefutable teacher, the sole, all-perfect Buddha." In the Pârâgika section of the Vinaya Pitaka, a conversation is recorded between Buddha

[1] *Legends and Theories of the Buddhists*, 1866, p. 171.

and a Brahman, who accused him of not honoring aged Brahmans, of not rising in their presence, and of not inviting them to be seated. Buddha replied, "Brahman, I do not see any one in the heavenly worlds nor in that of Mâra, nor among the inhabitants of the Brahma-worlds, nor among gods or men, whom it would be proper for me to honor, or in whose presence I ought to rise up, or whom I ought to request to be seated. Should the Tathâgata (Buddha) thus act towards any one, that person's head would fall off."

Such doctrines, as Gogerly points out, are irreconcilable with the doctrine of a universal Creator, who must necessarily be superior to all the beings formed and supported by him. But the most decisive passage on the subject is one taken from the Brahma-*gâla*-sûtra,[1] the first in the Dîrgha nikâya, which is itself the first work of the Sûtra Pitaka. It was translated by Gogerly, whose translation I follow, as the text has not yet been published. In the Brahma-*gâla*-sûtra, Buddha discourses respecting the sixty-two different sects; among whom four held the doctrine both of the preëxistence of the soul, and of its eternal duration through countless transmigrations. Others believed that some souls have always existed, whilst others have had a commencement of existence. Among these one sect is described as believing in the existence of a Creator, and it is here that Buddha brings together his arguments against the correctness of this opinion. "There is a time," he says, "O Bhikshus, when, after a very long period, this world is destroyed. On the destruction of the world very many beings ob-

---

[1] See J. D'Alwis's *Pâli Grammar*, p. 88, note; Turnour, *Mahâvansa*, Appendix iii. p. lxxv.

tained existence in the Âbhâsvara[1] Brahmaloka, which is the sixth in the series, and in which the term of life

[1] The Âbhâsvara gods, âbhâssarâ in Pâli, are mentioned already in the Dhammapada, v. 200, but none of the minute details, describing the six worlds of the gods, and the sixteen worlds of Brahman, and the four of Arûpa, are to be found there. The universe is represented (v. 126) as consisting of hell (niraya), earth, heaven (svarga), and Nirvâna. In v. 44 we find the world of Yama, the earth, and the world of the gods; in v. 104 we read of gods, Gandharvas, Mâra, and Brahman. The ordinary expression, too, which occurs in almost all languages, namely, in this world and in the next, is not avoided by the author of the Dhammapada. Thus we read in v. 168, "amim loke paramhi ka," in this world and in the next (cf. vv. 242, 410); we find in v. 20 "idhâ vâ huram và," here or there; in v. 15–18 we find "idha" and "pekka," here and yonder; pekka, i. e. pretya, meaning literally, "after having died" (cf. vv. 131, 306). We also find "idh'eva," here (v. 402), and "idha lokasmin," here in the world (v. 247), or simply "loke," in this world (v. 89); and "parattha" for "paratra," yonder, or in the other world.

A very characteristic expression, too, is that of v. 176, where, as one of the greatest crimes, is mentioned the scoffing at another world.

The following is a sketch of the universe and its numerous worlds, according to the later systems of the Buddhists. There are differences, however, in different schools.

1. The infernal regions:
    (1) Nyaya, hell.
    (2) The abode of animals.
    (3) The abode of Pretas, ghosts.
    (4) The abode of Asuras, demons.
2. The earth:
    (1) Abode of men.
3. The worlds of the gods:
    (1) Katur-mahârîga (duration, 9,000,000 years).
    (2) Trayastrimsa (duration, 36,000,000 years).
    (3) Yâma (duration, 144,000,000 years).
    (4) Tushita (duration, 576,000,000 years).
    (5) Nirmâna rati (duration, 2,304,000,000 years).
    (6) Paranirmita-vasavartin (duration, 9,216,000,000 years).
4. The worlds of Brahman:
    (a) First Dhyâna:
    (1) Brahma-parishadya (duration, ⅓ kalpa).
    (2) Brahma-purohita (duration, ½ kalpa).
    (3) Mahâbrahman (duration, one kalpa).

never exceeds eight kalpas. They are there spiritual beings (having purified bodies, uncontaminated with evil passions, or with any corporeal defilement); they have intellectual pleasures, are self-resplendent, traverse the atmosphere without impediment, and remain for a long time established in happiness. After a very long period this mundane system is reproduced, and the world named Brahma-vimâna (the third of the Brahmalokas) comes into existence, but uninhabited."

"At that time a being, in consequence either of the period of residence in Âbhâsvara being expired, or in consequence of some deficiency of merit preventing him from living there the full

    (*b*) Second Dhyâna:
  (4) Parîttâbha (duration, two kalpas).
  (5) Apramânâbha (duration, four kalpas).
  (6) Âbhâsvara (duration, eight kalpas).
    (*c*) Third Dhyâna:
  (7) Parîttasubha (duration, sixteen kalpas).
  (8) Apramânasubha (duration, thirty-two kalpas).
  (9) Subhakritsna (duration, sixty-four kalpas).
    (*d*) Fourth Dhyâna:
      (Anabhraka, of Northern Buddhism.)
      (Pu*n*ya-prasava, of Northern Buddhism.)
 (10) V*r*ihat-phala (500 kalpas).
 (11) Arangisattvas or Asangisattvas, of Nepal; **Asanyasatya,** of Ceylon (500 kalpas).
    (*e*) Fifth Dhyâna:
 (12) Av*r*iha (1,000 kalpas).
 (13) Atapa (2,000 kalpas).
 (14) Sud*r*isa (4,000 kalpas).
 (15) Sudarsana (8,000 kalpas).
      (Sumukha, of Nepal.)
 (16) Akanish*th*a (16,000 kalpas).
5. The world of Arûpa:
  (1) Âkâsânantyâyatanam (20,000 kalpas).
  (2) Vi*g*ñânânantyâyatanam (40,000 kalpas).
  (3) Akiñkanyâyatanam (60,000 kalpas).
  (4) Naivasañgnânâsañgnâyatanam (30,000 kalpas).
    Cf. Burnouf, *Introduction*, p. 599 *seq.*; *Lotus*, p. 811 *seq.*; **Hardy**, *Manual*, p. 25 *seq.*; Bigandet, p. 449.

period, ceased to exist in Âbhâsvara, and was reproduced in the uninhabited Brahma-vimâna. He was there a spiritual being; his pleasures were intellectual; he was self-resplendent, traversed the atmosphere, and, for a long time, enjoyed uninterrupted felicity. After living there a very long period in solitude, a desire of having an associate is felt by him, and he says, 'Would that another being were dwelling in this place.' At that precise juncture another being ceasing to exist in Âbhâsvara, comes into existence in the Brahma-vimâna, in the vicinity of the first one. They are both of them spiritual beings, have intellectual pleasures, are self-resplendent, traverse the atmosphere, and are, for a long time, in the enjoyment of happiness. Then the following thoughts arose in him who was the first existent in that Brahma-loka: 'I am Brahma, the Great Brahma, the Supreme, the Invincible, the Omniscient, the Governor of all things, the Lord of all. I am the Maker, the Creator of all things; I am the Chief, the disposer and controller of all, the Universal Father. This being was made by me. How does this appear? Formerly I thought, Would that another being were in this place, and upon my volition this being came here.' Those beings also, who afterwards obtained an existence there, thought, 'This illustrious Brahma is the Great Brahma, the Supreme, the Invincible, the Omniscient, the Ruler, the Lord, the Creator of all. He is the Chief, the Disposer of all things, the Controller of all, the Universal Father. We were created by him, for we see that he was first here, and that we have since then obtained existence. Furthermore, he who first obtained existence there lives during a very long period, exceeds in beauty, and is of immense power, but those who followed him are short-lived, of inferior beauty, and of little power.'

"It then happens, that one of those beings ceasing to exist there, is born in this world, and afterwards retires from society and becomes a recluse. He subjects his passions, is persevering in the practice of virtue, and by profound meditation he recollects his immediately previous state of existence, but none prior to that; he therefore says, 'That illustrious Brahma is the Great Brahma, the Supreme, the Invincible, the Omniscient, the Ruler, the Lord, the Maker, the Creator of all. He is the Chief, the Disposer of all things, the Controller of all, the Universal Father. That Brahma by whom we were created is ever-enduring, immutable, eternal, unchangeable, continuing forever the same.

But we, who have been created by this illustrious Brahma, are mutable, short-lived, and mortal.'"

There is, it seems to me, an unmistakable note of irony in this argumentation against the belief in a personal Creator; and to any one acquainted with the language of the Upanishads, the pointed allusions to expressions occurring in those philosophical and religious treatises of the Brahmans are not to be mistaken. If then it is true, as Gogerly remarks, that many who call themselves Buddhists acknowledge the existence of a Creator, the question naturally arises, whether the point-blank atheism of the Brahma-*g*âla was the doctrine of the founder of Buddhism or not.

This is, in fact, but part of the problem so often started, whether it is possible to distinguish between Buddhism and the personal teaching of Buddha. We possess the Buddhist Canon, and whatever is found in that canon we have a right to consider as the orthodox Buddhist doctrine. But as there has been no lack of efforts in Christian theology to distinguish between the doctrine of the founder of our religion and that of the writers of the Gospels, to go beyond the canon of the New Testament, and to make the λόγια of the Master the only solid rule of our faith, so the same want was felt at a very early period among the followers of Buddha. King Asoka, the Indian Constantine, had to remind the assembled priests at the great council which had to settle the Buddhist Canon, that, "*what had been said by Buddha, that alone was well said.*"[1] Works attributed to Buddha, but declared to be apocryphal, or even heterodox, existed already at that time (246 B. C.). Thus we are by no means without authority for distinguishing between Buddhism and the

---

[1] M. M.'s *Chips from a German Workshop*, 2d ed. vol. i. p. xxiv.

teaching of Buddha; the only question is, whether in our time such a separation is still practicable?

My belief is that, in general, all honest inquirers must oppose a No to this question, and confess that it is useless to try to cast a glance beyond the boundaries of the Buddhist Canon. What we find in the canonical books in the so-called "Three Baskets," is orthodox Buddhism and the doctrine of Buddha, similarly as we must accept in general whatever we find in the four Gospels as orthodox Christianity and the doctrine of Christ.

Still, with regard to certain doctrines and facts, the question, I think, ought to be asked again and again whether it may not be possible to advance a step further even with the conviction that we cannot arrive at results of apodictic certainty? If it happens that on certain points we find in different parts of the canon, not only doctrines differing from each other, but plainly contradictory to each other, it follows, surely, that one only of these can have belonged to Buddha personally. In such a case, therefore, I believe we have a right to choose, and I believe we shall be justified in accepting that view as the original one, the one peculiar to Buddha himself, which harmonizes *least* with the later system of orthodox Buddhism.

As regards the denial of a Creator, or atheism in the ordinary acceptation of the word, I do not think that any one passage from the books of the canon known to us can be quoted which contravenes it, or which in any way presupposes the belief in a personal God or Creator. All that might be urged are the words said to have been spoken by Buddha at the time when he became the Enlightened, the Buddha. They are as follows: "Without ceasing shall I run through a course of many births, looking for the maker of this tabernacle,—and painful is birth again and again. But now, maker of the tabernacle, thou hast been seen; thou shalt not make up this tabernacle again. All thy rafters are broken, thy ridge-pole is sundered; the mind, being sundered, has attained to the extinction of all desires."

Here in the maker of the tabernacle, *i. e.*, the body, one might be tempted to see a creator. But he who is acquainted with the general run of thought in Buddhism, soon finds that this architect of the house is only a poetical expression, and that whatever meaning may underlie it, it evidently signifies a force subordinate to the Buddha, the Enlightened.

But whilst we have no ground for exonerating the Buddha personally from the accusation of atheism, the matter stands very differently as regards the charge of nihilism. The Buddhist nihilism has always been much more incomprehensible than mere atheism. A kind of religion is still conceivable, when there is something firm somewhere, when a something, eternal and self-dependent, is recognized, if not *without* and *above* man, at least *within* him. But if, as Buddhism teaches, the soul, after having passed through all the phases of existence, all the worlds of the gods and of the higher spirits, attains finally Nirvâna as its highest aim and last reward, *i. e.* becomes utterly extinct, then religion is not any more what it is meant to be,—a bridge from the finite to the infinite, but a trap-bridge hurling man into the abyss at the very moment when he thought he had arrived at the stronghold of the Eternal. According to the metaphysical doctrine of Buddhism, the soul cannot dissolve itself in a higher being, or be absorbed in the absolute substance, as was taught by the Brahmans, and other mystics of ancient and modern times; for Buddhism knew not the Divine, the Eternal, the Absolute; and the soul even as the I, or as the mere Self, the Âtman, as called by the Brahmans, was represented in the orthodox metaphysics of Buddhism as transient, as futile, as a mere phantom.

No person who reads with attention the metaphysical speculations on the Nirvâna contained in the third part of the Buddhist Canon, can arrive at any other conviction than that expressed by Burnouf, namely, that Nirvâna, the highest aim, the *summum bonum* of Buddhism, is the absolute nothing.

Burnouf adds, however, that this doctrine appears in its crude form in the third part only of the canon, the so-called Abhidharma, but not in the first and second parts, in the Sûtras, the sermons, and the Vinaya, the ethics, which together bear the name of Dharma, or Law. He next points out that, according to some ancient authorities, this entire part of the canon was designated as not "pronounced by Buddha."[1] These are, at once, two important limitations. I add a third, and maintain that sayings of Buddha occur in the Dhammapada, which are in open contradiction to this metaphysical nihilism.

Now, first, as regards the soul, or the self, the existence of which, according to the orthodox metaphysics, is purely phenomenal,[2] a sentence attributed to the Buddha[3] says, "Self is the Lord of Self, who else

---

[1] Max Müller's *Chips*, 2d ed. vol. i. p. 285, note.

[2] See Wassiljew, *Der Buddhismus*, p. 296, (269); and Bigandet's *Life of Gaudama*, p. 479. "The things that I see and know, are not myself, nor from myself, nor to myself. What seems to be myself is in reality neither myself nor belongs to myself."

[3] *Dhammapada*, v. 160.

could be the Lord?" And again,[1] "A man who controls himself enters the untrodden land through his own self-controlled self." But this untrodden land is the Nirvâna.

Nirvâna certainly means extinction, whatever its later arbitrary interpretations[2] may have been, and seems therefore to imply, even etymologically, a real blowing out or passing away. But Nirvâna occurs also in the Brahmanic writings as synonymous with Moksha,[3] Nirvritti,[3] and other words, all designating the highest stage of spiritual liberty and bliss, but not annihilation. Nirvâna may mean the extinction of many things, — of selfishness, desire, and sin, without going so far as the extinction of subjective consciousness. Further, if we consider that Buddha himself, after he had already seen Nirvâna, still remains on earth until his body falls a prey to death; that in the legends Buddha appears to his disciples even after his death, it seems to me that all these circumstances are hardly reconcilable with the orthodox metaphysical doctrine of Nirvâna.

But I go even further and maintain that, if we look in the Dhammapada at every passage where Nirvâna is mentioned, there is not one which would require that its

---

[1] *Dhammapada*, v. 323.

[2] See Bastian, *Die Völker des östlichen Asien*, vol. iii. p. 354. The learned abbot who explained the meaning of Nirvâna to Dr. Bastian was well versed in the old grammatical terminology. He distinguishes the causal meaning, called hetumat, of the verb "vâ," to cause to blow out, from the intransitive meaning, to go out. He also distinguishes between the verb as expressing the state of vanishing, "bhâvasâdhana" (cf. *Pân.* ii. 3, 37; iii. 4, 69), or the place of vanishing, "adhikaranasâdhana" (*Pân.* i. 4, 45). How place and act become one in the conception of Buddhists, is better seen by the four dhyânas, originally meditations, than the places reached by these meditations.

[3] See *Dhammapada*, v. 89, 92.

meaning should be annihilation, while most, if not all, would become perfectly unintelligible if we assigned to the word Nirvâ*n*a the meaning which it has in the Abhidharma or the metaphysical portions of the canon.

What does it mean, when Buddha (v. 21), calls reflection the path to immortality, thoughtlessness the path of death? Buddhaghosha does not hesitate to explain immortality by Nirvâna, and that the same idea was connected with it in the mind of Buddha is clearly proved by a passage immediately following (v. 23): "The wise people, meditative, steady, always possessed of strong powers, attain to Nirvâna, the highest happiness." In the last verse, too, of the same chapter we read, " A Bhikshu who delights in reflection, who looks with fear on thoughtlessness, will not go to destruction, — he is near to Nirvâna." If the goal at which the followers of Buddha have to aim had been in the mind of Buddha perfect annihilation, "amata," *i. e.* immortality, would have been the very last word he could have chosen as its name.

In several passages of the Dhammapada, Nirvâ*n*a occurs in the purely ethical sense of rest, quietness, absence of passion; *e. g.* (v. 134), " If, like a trumpet trampled under foot, thou utter not, then thou hast reached Nirvâna; anger is not known in thee." In v. 184 long-suffering (titikshâ) is called the highest Nirvâna. While in v. 202 we read that there is no happiness like rest (sânti) or quietness, we read in the next verse that the highest happiness is Nirvâna. In v. 285, too, " sânti " seems to be synonymous with Nirvâna, for the way that leads to " sânti," or peace, leads also to Nirvâna, as shown by Buddha. In v. 369 it is said, " When thou hast cut off passion and hatred, thou wilt go to Nirvâ*n*a;" and in v. 225 the same thought is expressed, only that instead of Nirvâna we have the expression of unchangeable place: "The sages who injure nobody, and who always control their body, they will go to the unchangeable place, where, if they have gone, they will suffer no more."

In other passages Nirvâna is described as the result of right knowledge. Thus we read (v. 203), "Hunger is the worst of diseases, the body the greatest of pains; if one knows this truly, that is Nirvâna, the highest happiness."

A similar thought seems contained in v. 374: "As soon as a man has perceived the origin and destruction of the elements of the body (khandha), he finds happiness and joy, which belong to those who know the immortal (Nirvâna); or which is the immortality of those who know it, namely, the transitory character of the body." In v. 372 it is said that he who has knowledge and meditation is near unto Nirvâna.

Nirvâna is certainly more than heaven or heavenly joy. "Some people are born again" (on earth), says Buddha, v. 126, "evil-doers go to hell; righteous people go to heaven; those who are free from all worldly desires enter Nirvâna." The idea that those who had reached the haven of the gods were still liable to birth and death, and that there is a higher state in which the power of birth and death is broken, existed clearly at the time when the verses of the Dhammapada were composed. Thus we read (v. 238), "When thy impurities are blown away, and thou art free from guilt, thou wilt not enter again into birth and decay." And in the last verse the highest state that a Brâhmana can reach is called "the end of births," *gâti-kshaya*.

There are many passages in the Dhammapada where

we expect Nirvâna, but where, instead of it, other words are used. Here, no doubt, it might be said that something different from Nirvâna is intended, and that we have no right to use such words as throwing light on the original meaning of Nirvâna. But, on the other hand, these words, and the passages where they occur, must mean something definite; they cannot mean heaven or the world of the gods, for reasons stated above; and if they do not mean Nirvâna, they would have no meaning at all. There may be some doubt whether " pâra," the shore, and particularly the other shore, stands always for Nirvâna, and whether those who are said to have reached the other shore, are to be supposed to have entered Nirvâna. It may possibly not have that meaning in verses 384 and 385, but it can hardly have another in places such as vv. 85, 86, 347, 348, 355, 414. There is less doubt, however, that other words are used distinctly as synonyms of Nirvâna. Such words are, the quiet place (sântam padam, vv. 368, 381); the changeless place (akyutam sthânam, v. 225, compared with v. 226); the immortal place (amatam padam, v. 114); also simply that which is immortal (v. 374). In v. 411 the expression occurs that the wise dives into the immortal.

Though, according to Buddha, everything that has been made, everything that was put together, resolves itself again into its component parts and passes away (v. 277, sarve samskârâ anityâh), he speaks nevertheless of that which is not made, *i. e.*, the uncreated and eternal, and uses it, as it would seem, synonymously with Nirvâna (v. 97). Nay, he says (v. 383), " When you have understood the destruction of all that was made, you will understand that which was not made." This surely shows that even for Buddha a something existed which is not made, and which, therefore, is imperishable and eternal.

On considering such sayings, to which many more might be added, one recognizes in them a conception of Nirvana, altogether irreconcilable with the nihilism of the third part of the Buddhist Canon. It is not a question of more or less, but of *aut—aut.* Nirvâna cannot, in the mind of one and the same person, mean black and white, nothing and something. If these sayings, as recorded in the Dhammapada, have maintained themselves, in spite of their being in open contradiction to orthodox metaphysics, the only explanation, in my opinion, is, that they were too firmly fixed in the tradition which went back to Buddha and his disciples. What Bishop Bigandet and others represent as the popular view of Nirvâna, in contradistinction to that of the Buddhist divines, was, in my opinion, the conception of Buddha and his disciples. It represented the entrance of the soul into rest, a subduing of all wishes and desires, indifference to joy and pain, to good and evil, an absorption of the soul in itself, and a freedom from the circle of existences from birth to death, and from death to a new birth. This is still the meaning which educated people attach to it, whilst to the minds of the larger masses [1] Nirvâna suggests rather the idea of a Mohammedan paradise or of blissful Elysian fields.

Only in the hands of the philosophers, to whom Buddhism owes its metaphysics, the Nirvâna, through constant negations carried to an indefinite degree, through the excluding and abstracting of all that is not Nirvâna, at last became an empty Nothing, a phi-

---

[1] Bigandet, *The Life of Gaudama,* p. 320, note; Bastian, *Die Völger des östlichen Asien,* vol. iii. p. 353.

losophical myth. There is no lack of such philosophical myths either in the East or in the West. What has been fabled by philosophers of a Nothing, and of the terrors of a Nothing, is as much a myth as the myth of Eos and Tithonus. There is no more a Nothing than there is an Eos or a Chaos. All these are sickly, dying, or dead words, which, like shadows and ghosts, continue to haunt language, and succeed in deceiving for a while even the healthiest intellect.

Even modern philosophy is not afraid to say that there is a Nothing. We find passages in the German mystics, such as Eckhart and Tauler, where the abyss of the Nothing is spoken of quite in a Buddhist style. If Buddha had said, like St. Paul, " that what no eye hath seen, nor ear heard, neither has it entered into the heart of man," was prepared in the Nirvâna for those who had advanced to the highest degree of spiritual perfection, such expressions would have been quite sufficient to serve as a proof to the philosophers by profession that this Nirvâna, which could not become an object of perception by the senses, nor of conception by the categories of the understanding, — the anâkkhâta, the ineffable, as Buddha calls it (v. 218),— could be nothing more nor less than the Nothing. Could we dare with Hegel to distinguish between a Nothing (*Nichts*) and a Not (*Nicht*), we might say that the Nirvâna had, through a false dialectical process, been driven from a relative Nothing to an absolute Not. This was the work of the theologians and of the orthodox philosophers. But a religion has never been founded by such teaching, and a man like Buddha, who knew mankind, must have known that he could not, with such weapons, overturn the tyranny of the Brahmans. Either we must bring ourselves to believe that Buddha taught his disciples two diametrically opposed doctrines on Nirvâna, say an exoteric and esoteric one, or we must allow *that* view of Nirvâna to have been the original view of the founder of this marvelous religion, which we find recorded in the verses of the Dhammapada, and which corresponds best with the simple, clear, and practical character of Buddha.

ON THE TITLE OF THE DHAMMAPADA.

I have still to say a few words on the title of the Dhammapada. This title was first rendered by Gogerly, " The Footsteps of Religion; " by Spence Hardy, " The Paths of Religion," and this, I believe, is in the main a correct rendering. "Dharma," or, in Pâli, "dhamma," has many meanings. Under one aspect, it means religion, in so far, namely, as religion is the law that is to be accepted and observed. Under another aspect " dharma " is virtue, in so far, namely, as virtue is the realization of that law. Thus " dharma " can be rendered by law, by religion, more particularly Buddha's religion, or by virtue.

"Pada," again, may be rendered by footsteps, but its more natural rendering is path. Thus we read in verse 21, "appamâdo amatapadam," reflection is the path of immortality, *i. e.*, the path that leads to immortality. Again, " pamâdo makkuno padam," thoughtless is the path of death, *i. e.*, the path that leads to death. The commentator explains " padam " here by "amatasya adhigamupâya," the means of obtaining immortality, *i. e.*, Nirvâna, or simply by " upâyo " and " magga," the way.[1] In the same manner " dhamma-

---

[1] If we compare verses 92 and 93, and again 254 and 255, we see that " padam " is used synonymously with " gati," going.

padam" would mean "the path of virtue," *i. e.*, the path that leads to virtue, a very appropriate title for a collection of moral precepts. In this sense "dhammapadam" is used in verses 44 and 45, as I have explained in my notes to these verses.

Gogerly, though not to be trusted in all his translations, may generally be taken as a faithful representative of the tradition of the Buddhists in Ceylon, and we may therefore take it for granted that the priests of that island take Dhammapada to mean, as Gogerly translates it, the vestiges of religion, or, from a different point of view, the path of virtue.

It is well known, however, that the learned editor of the Dhammapada, Dr. Fausböll, proposed a different rendering. On the strength of verses 44 and 102, he translated "dhammapada" by "collection of verses on religion." But though "pada" may mean a verse, I doubt whether "pada" in the singular could ever mean a collection of verses. In verse 44 "padam" cannot mean a collection of verses, for reasons I have explained in my notes; and in verse 102 we have, it seems to me, the best proof that, in Buddhist phraseology, "dhammapada" is not to be taken in a collective sense, but means a law-verse, a wise saw. For there we read, "Though a man recite a hundred Gâthâs made up of senseless words, one 'dhammapada,' *i. e.*, one single word or line of the law, is better, which if a man hears, he becomes quiet." If the Buddhists wish to speak of many law-verses, they use the plural, dhammapadâni.[1] Thus Buddhaghosha says,[2] "Be it known that the Gâthâ consists of the Dhammapadâni. Theragâthâ, Therîgâthâ, and those un-

---

[1] "Pada" by itself forms the plural "padâ," as in v. 243, *k*aturo padâ.

[2] D'Alwis, *Pâli Grammar*, p. 61.

mixed (detached) Gâthâ not comprehended in any of the above-named Suttantâ."

Unless, therefore, it can be proved that in Pâli, "padam" in the singular can be used in a collective sense, so as to mean a collection of words or sayings, and this has never been done, it seems to me that we must retain the translation of Gogerly, "Footsteps of Religion," though we may with advantage make it more intelligible in English by rendering it "The Path of Virtue." The idea of representing life, and particularly the life of the faithful, as a path of duty or virtue leading to deliverance (in Sanskrit, dharmapatha) is very familiar to the Buddhists. The four great truths [1] of their religion consist in the recognition, (1) that there is suffering; (2) that there is a cause of that suffering; (3) that such cause can be removed; (4) that there is a way of deliverance, namely, the doctrine of Buddha. This way, this mârga, is then fully described as consisting of eight stations,[2] and leading in the end to Nirvâna.[3] The faithful advances on that road, 'padât padam,' step by step, and it is therefore called patipadâ, lit. the step by step.[4]

The only way in which Dhammapadam could possibly be defended in the sense of "Collection of verses

---

[1] Spence Hardy, *Manual*, p. 496.   [2] Ibid.

[3] Burnouf, *Lotus*, p. 520. "Ajoutons, pour terminer ce que nous trouvons à dire sur le mot *magga*, quelque commentaire qu'on en donne d'ailleurs, que suivant une définition rapportée par Turnour, le *magga* renferme une sous-division que l'on nomme *patipadâ*, en sanscrit *pratipad*. Le *magga*, dit Turnour, est la voie qui conduit au Nibbâna, le *patipadâ*, littéralement 'la marche pas à pas, ou le degré,' est la vie de rectitude qu'on doit suivre, quand on marche dans la voie de *magga*."

[4] See Spence Hardy, *Manual*, p. 496. Should not "*katurvidha-dharma-pada*," mentioned on p. 497, be translated by "the fourfold path of the Law?" It can hardly be the fourfold word of the Law.

of the Law," would be if we took it for an aggregate compound. But such aggregate compounds, in Sanskrit at least, are possibly only with numerals, as, for instance, Tri-bhuvanam, the three worlds, *katuryugam*, the four ages.[1] It might, therefore, be possible to form in Pâli also such compounds as dasapadam, a collection of ten padas, a work consisting of ten padas, a "decamerone"; but it would in no way follow that we could attempt such a compound as Dhammapadam, in the sense of collection of law-verses.

I find that Dr. Köppen has been too cautious to adopt Dr. Fausböll's rendering, while Professor Weber, of Berlin, not only adopts that rendering without any misgivings, but in his usual way blames me for my backwardness.[2]

In conclusion, I have to say a few words on the spelling of technical terms which occur in the translation of the Dhammapada and in my introduction. It is very difficult to come to a decision on this subject; and I have to confess that I have not been consistent throughout in following the rule which I think ought to be followed. Most of the technical terms employed by Buddhist writers come from Sanskrit; and in the eyes of the philologist the various forms which they have assumed in Pâli, in Burmese, in Tibetan, in Chinese, in Mongolian, are only so many corruptions

---

[1] See M. M.'s *Sanskrit Grammar*, § 519.

[2] "Dies ist eben auch der Sinn, der dem Titel unseres Werkes zu geben ist (nicht, 'Footsteps of the Law' wie *neuerdings noch* M. Müller will. s. dessen *Chips from a German Workshop*, vol. i. p. 200). The fact is that on page 200 of my *Chips* there is no mention of the Dhammapada at all, while on page 220 I had simply quoted from Spence Hardy, and given the translation of Dhammapada, "Footsteps of the Law," between inverted commas.

of the same original form. Everything, therefore, would seem to be in favor of retaining the Sanskrit forms throughout, and of writing, for instance, Nirvâna instead of the Pâli Nibbâna, the Burmese Niban or Nepbhân, the Siamese Niruphan, the Chinese Nipan. The only hope, in fact, that writers on Buddhism will ever arrive at a uniform and generally intelligible phraseology seems to lie in their agreeing to use throughout the Sanskrit terms in their original form, instead of the various local disguises and disfigurements which they present in Ceylon, Burmah, Siam, Tibet, China, and Mongolia. But against this view another consideration is sure to be urged, namely, that many Buddhist words have assumed such a strongly marked local or national character in the different countries and in the different languages in which the religion of Buddha has found a new home, that to translate them back into Sanskrit would seem as affected, nay, prove in certain cases as misleading, as if, in speaking of *priests* and *kings*, we were to speak of *presbyters* and *cynings*. Between the two alternatives of using the original Sanskrit forms or adopting their various local varieties, it is sometimes difficult to choose, and the rule by which I have been mainly guided has been to use the Sanskrit forms as much as possible; in fact, everywhere except where it seemed affected to do so. I have therefore written Buddhaghosha instead of the Pâli Buddhaghosa, because the name of that famous theologian, "the Voice of Buddha," seemed to lose its significance if turned into Buddhaghosa. But I am well aware what may be said on the other side. The name of Buddhaghosha, "Voice of Buddha," was given him after he had been converted from Brahmanism to Buddhism, and it was given to him by people to whom the Pâli word *ghosa* conveyed the same meaning as *ghosha* does to us. On the other hand, I have retained the Pâli *Dhammapada* instead of Dharmapada, simply because, as the title of a Pâli book, it has become so familiar that to speak of it as Dharmapada seemed like speaking of another work. We are accustomed to speak of Samanas instead of *S*ramanas, for even in the days of Alexander's conquest, the Sanskrit word *S*rama*n*a had assumed the prakritized or vulgar form which we find in Pâli, and which alone could have been rendered by the later Greek writers (first by Alexander Polyhistor, 80–60, B. C.) by σαμαναῖοι.[1] As a Buddhist term, the Pâli form Samana has so entirely supplanted that of *S*rama*n*a that, even in the Dhammapada (v. 388) we find an etymology of Samana as derived from " sam," to be quiet, and not from " sram," to toil. But though one might bring oneself to speak of Samanas, who would like to introduce Bâhma*n*a instead of Brâhma*n*a? And yet this word, too, had so entirely been replaced by bâhmana, that in the Dhammapada, it is derived from a root " vah," to remove, to separate, to cleanse.[2] My own conviction is that it would be best if writers on Buddhist literature and religion were to adopt Sanskrit throughout as the *lingua franca*. For an accurate understanding of the original meaning of most of the

---

[1] See Lassen, *Indische Alterthumskunde*, vol. ii. p. 700, note. That Lassen is right in taking the Σαρμάναι, mentioned by Megasthenes, for Brahmanic, not for Buddhist ascetics, might be proved also by their dress. Dresses made of the bark of trees are not Buddhistic. On page lxxix., note, read Alexander Polyhistor instead of Bardesanes.

[2] See *Dhammapada*, v. 388; Bastian, *Völker des östlichen Asien*, vol. iii. p. 412 : " Ein buddhistischer Mönch erklärte mir, dass die Brahmanen ihren Namen führten, als Leute, die ihre Sünden abgespült hätten." See also *Lalita-vistara*, p. 551, line 1 ; p. 553, line 7.

technical terms of Buddhism a knowledge of their Sanskrit form is indispensable; and nothing is lost, while much would be gained, if, even in the treating of Southern Buddhism, we were to speak of the town of Srâvasti instead of Sâvatthi in Pâli, Sevet in Singhalese; of Tripitaka, "the three baskets," instead of Pitakattaya in Pâli, Tunpitaka in Singhalese; of Arthakathâ, "commentary,' instead of Atthakathâ in Pâli, Atuwâva in Singhalese; and therefore also of Dharmapada, "the path of virtue," instead of Dhammapada.

MAX MÜLLER.

DUSTERNBROOK, near KIEL, in the summer of 1869.

CHAPTER I.

THE TWIN-VERSES.

1.

ALL that we are is the result of what we have thought: it is founded on our thoughts, it is made up of our thoughts. If a man speaks or acts with an evil thought, pain follows him, as the wheel follows the foot of him who draws the carriage.

2.

All that we are is the result of what we have thought: it is founded on our thoughts, it is made up of our thoughts. If a man speaks or acts with a pure thought, happiness follows him, like a shadow that never leaves him.

(1.) "Dharma," though clear in its meaning, is difficult to translate. It has different meanings in different systems of philosophy, and its peculiar application in the phraseology of Buddhism has been fully elucidated by Burnouf, *Introduction à l'Histoire du Bud-dhisme*, p. 41 *seq.* He writes: "Je traduis ordinairement ce terme par *condition*, d'autres fois par *lois*, mais aucune de ces traductions n'est parfaitement complète; il faut entendre par 'dharma' ce qui fait qu'une chose est, ce qu'elle est, ce qui constitue sa nature propre, comme l'a bien montré Lassen, à l'occasion de la célèbre formule, 'Ye dharmâ hetuprabhavâ.'" Etymologically the Latin *for-ma* expresses the same general idea which was expressed by "dhar-ma." See also Burnouf, *Lotus de la bonne Loi*, p. 524. Fausböll translates: "Naturæ a mente principium ducunt," which shows that he understood "dharma" in the Buddhist sense. Gogerly and D'Alwis translate: "Mind precedes action," which, if not wrong, is at all events wrongly expressed; while Professor Weber's rendering, "Die Pflichten aus dem Herz folgern," is quite inadmissible.

3.

"He abused me, he beat me, he defeated me, he robbed me,"—hatred in those who harbor such thoughts will never cease.

4.

"He abused me, he beat me, he defeated me, he robbed me,"—hatred in those who do not harbor such thoughts will cease.

5.

For hatred does not cease by hatred at any time: hatred ceases by love; this is an old rule.

6.

And some do not know that we must all come to an end here; but others know it, and hence their quarrels cease.

7.

He who lives looking for pleasures only, his senses uncontrolled, immoderate in his enjoyments, idle, and

(3.) On "akko*kkh*i," see Ka*kkâ*yana, vi. 4, 17. D'Alwis, *Pâli Grammar*, p. 38, note. "When akko*kkh*i means 'abused,' it is derived from 'kunsa,' not from 'kudha.'"

(6.) It is necessary to render this verse freely, because literally translated it would be unintelligible. "Pare" is explained by fools, but it has that meaning by implication only. There is an opposition between "pare *ka*" and "ye *ka*," which I have rendered by "some" and "others." Yamâmase, a 1 pers. plur. imp. âtm., but really a *Let* in Pâli See Fausböll, *Five Gâtakas*, p. 38.

(7.) "Mâra" must be taken in the Buddhist sense of tempter, or evil spirit. See Burnouf, *Introduction*, p. 76: "Mâra est le démon de l'amour, du péché et de la mort; c'est le tentateur et l'ennemi de Buddha." As to the definite meaning of "vîrya," see Burnouf, *Lotus*, p. 548.

"Kusîta," idle, is evidently the Pâli representative of the Sanskrit "kusîda." In Sanskrit "kusîda," slothful, is supposed to be derived from "sad," to sit, and even in its other sense, namely, a loan, it may have been intended originally for a pawn, or something that lies weak, Mâra (the tempter) will certainly overcome him, as the wind throws down a weak tree.

inert. In the Buddhistical Sanskrit, "kusîda" is the exact counterpart of the Pâli "kusîta;" see Burnouf, *Lotus*, p. 548. But supposing "kusîda" to be derived from "sad," the d would be organic, and its phonetic change to t in Pâli, against all rules. I do not know of any instance where an original Sanskrit d, between two vowels, is changed to t in Pâli. The Pâli "dandham" (*Dhammap.* v. 116) has been identified with "tandram," lazy; but here the etymology is doubtful, and "dandra" may really be a more correct dialectic variety, *i. e.*, an intensive form of a root "dram" (dru) or "drâ." Anyhow the change here affects an initial, not a medial d, and it is supposed to be a change of Sanskrit t to Pâli d, not *vice versâ*. Professor Weber supposed "pithîyati" in v. 173, to stand for Sk. "pidhîyate," which is impossible. See Kakkayana's *Grammar*, iv. 21. Dr. Fausböll had identified it rightly with Sk. "apistîryati." Comparisons such as Pâli "alâpu" (v. 149) with Sk. "alâbu," and Pâli "pabbaga" (v. 345) with Sk. "bâlbaga," prove nothing whatever as to a possible change of Sk. d to Pâli t, for they refer to words the organic form of which is doubtful, and to labials instead of dentals.

A much better instance was pointed out to me by Mr. R. C. Childers, namely the Pâli "pâtu," Sk. "prâdus," clearly, openly. Here however, the question arises, whether "pitu" may not be due to dialectic variety, instead of phonetic decay. If "pâtu" is connected with "pràtar," before, early, "prâdus" would be a peculiar Sanskrit corruption, due to a mistaken recollection of "dus," while the Pâli "pâtu" would have preserved the original t.

Anyhow, we require far stronger evidence before we can admit a medial t in Pâli as a phonetic corruption of a medial d in Sanskrit. We might as well treat the O. H. G. t as a phonetic corruption of Gothic d. The only way to account for the Pâli form "kusîta" instead of "kusîda," is by admitting the influence of popular etymology. Pâli has in many case lost its etymological consciousness. It derives "samana" from a root "sam," "b (r) âhmana" from "bâh;" see v. 388. Now as "sîta" in Pâli means cold, apathetic, but in a good sense, "kusîta" may have been formed in Pâli to express apathetic in a bad sense.

Further, we must bear in mind that the Sanskrit etymology of "kusîda" from "sad," though plausible, is by no means certain. If, on the one hand, "kusîda" might have been misinterpreted in Pâli, and changed to "kusîta," it is equally possible that "kusîta," supposing this to have been the original form, was misinterpreted in Sanskrit, and changed there to "kusîda." "Sai" is mentioned as a

8.

He who lives without looking for pleasures, his senses well controlled, in his enjoyments moderate, faithful and strong, Mâra will certainly not overcome him, any more than the wind throws down a rocky mountain.

9.

He who wishes to put on the sacred orange-colored dress without having cleansed himself from sin, who

Sk. root in the sense of *tabescere;* from it "kusîta" might possibly be derived in the sense of idle. "Sita" in Sanskrit is what is sown, "sítà," the furrow; from it "kusîta" might mean a bad laborer. These are merely conjectures, but it is certainly remarkable that there is an old Vedic proper name Kushita-ka, the founder of the Kaushîtakas, whose Brâhma*n*a, the Kaushîtaki-brâhma*n*a, belongs to the Rig-Veda. An extract from it was translated in my *History of Ancient Sanskrit Literature*, p. 407.

Lastly, it should be mentioned, that while "kusîta" is the Pâli counterpart of "kusîda," the abstract name in Pâli is "kosa*gg*a," Sanskrit "kausîdya," and not "kosa*kk*a," as it would have been if derived from "kusîta."

(9.) The saffron dress, of a reddish-yellow or orange color, the Kâsâva or Kâshâya, is the distinctive garment of the Buddhist priests. The play on the words "anikkasâvo kâsâvam," or in Sanskrit, "anishkashâya*h* kâshâyam," cannot be rendered in English. "Kashâya" means, impurity, "nish-kashâya," free from impurity, "anishkashâya," not free from impurity, while "kâshâya" is the name of the orange-colored or yellowish Buddhist garment. The pun is evidently a favorite one, for, as Fausböll shows, it occurs also in the *Mahâbhârata*, xii. 568:—

> Anishkashâye kashâyam ihârtham iti viddhi tam,
> Dharmadhvagânâ*m* mun*d*ânâ*m* vrittyartham iti me mati*h*.

"Know that this orange-colored garment on a man who is not free from impurity, serves only for the purpose of cupidity; my opinion is, that it is meant to supply the means of living to those men with shaven heads, who carry their virtue like a flag."

(I read "vrittyartham," according to the Bombay edition, instead of "k*r*itartham," the reading of the Calcutta edition.)

With regard to "sîla," virtue, see Burnouf, *Lotus*, p. 547.

On the exact color of the dress, see Bishop Bigandet, *The Life or Legend of Gaudama, the Buddha of the Burmese*, Rangoon, 1866, p. 504.

disregards also temperance and truth, is unworthy of the orange-colored dress.

10.

But he who has cleansed himself from sin, is well grounded in all virtues, and regards also temperance and truth, is indeed worthy of the orange-colored dress.

11.

They who imagine truth in untruth, and see untruth in truth, never arrive at truth, but follow vain desires.

12.

They who know truth in truth, and untruth in untruth, arrive at truth, and follow true desires.

13.

As rain breaks through an ill-thatched house, passion will break through an unreflecting mind.

14.

As rain does not break through a well-thatched house, passion will not break through a well-reflecting mind.

15.

The evil-doer mourns in this world, and he mourns

(11, 12.) "Sara," which I have translated by truth, has many meanings in Sanskrit. It means the sap of a thing, then essence or reality; in a metaphysical sense, the highest reality; in a moral sense, truth. It is impossible in a translation to do more than indicate the meaning of such words, and in order to understand them fully, we must know not only their definition, but their history.

(15.) "Kili*tth*a" is "klish*t*a," a participle of "klis." It means literally, what is spoilt. The abstract noun "klesa," evil or sin, is constantly employed in Buddhist works; see Burnouf, *Lotus*, p. 443. Possibly the words were intended to be separated, "kamma kili*tt*ham," and not to be joined like "kamma-visuddhim" in the next verse.

in the next; he mourns in both. He mourns, he suffers when he sees the evil of his own work.

16.

The virtuous man delights in this world, and he delights in the next; he delights in both. He delights, he rejoices, when he sees the purity of his own work.

17.

The evil-doer suffers in this world, and he suffers in the next; he suffers in both. He suffers when he thinks of the evil he has done; he suffers more when going on the evil path.

18.

The virtuous man is happy in this world, and he is happy in the next; he is happy in both. He is happy when he thinks of the good he has done; he is still more happy when going on the good path.

19.

The thoughtless man, even if he can recite a large portion (of the law), but is not a doer of it, has no share in the priesthood, but is like a cowherd counting the cows of others.

(16.) Like "klish*t*a" in the preceding verse, "visuddhi" in the present has a technical meaning One of Buddhaghosha's most famous works is called "Visuddhi magga." See Burnouf, *Lotus*, p. 844.

(17, 18.) "The evil path and the good path" are technical expressions for the descending and ascending scale of worlds through which all beings have to travel upward or downward, according to their deeds. See Bigandet, *Life of Gaudama*, p. 5, note 4, and p. 449; Burnouf, *Introduction*, p. 599; *Lotus*, p. 865, l. 7; l. 11.

(19.) In taking "sahitam" in the sense of "samhitam" or "samhita," I follow the commentator, who says, "Tepi*t*akassa Buddavakanass*

20.

The follower of the law, even if he can recite only a small portion (of the law), but, having forsaken passion and hatred and foolishness, possesses true knowledge and serenity of mind, he, caring for nothing in this world, or that to come, has indeed a share in the priesthood.

etam nâmam," but I cannot find another passage where the Tripi*t*aka, or any portion of it, is called Sahita. "Sa*m*hita" in vv. 100-102, has a different meaning. The fact that some followers of Buddha were allowed to learn short portions only of the sacred writings by heart, and to repeat them, while others had to learn a larger collection, is shown by the story of *K*akkhupâla, p. 3, of Mahâkâla, p. 26, etc.

"Sama*n*na," which I have rendered by "priesthood," expresses all that belongs to, or constitutes a real sama*n*a or *s*rama*n*a, this being the Buddhist name corresponding to the bràhma*n*a, or priest, of the orthodox Hindus. Buddha himself is frequently called the Good Samana. Fausböll takes the abstract word "sâmanna" as corresponding to the Sanskrit "sâmânya," community, but Weber has well shown that it ought to be taken as representing "srâma*n*ya." He might have quoted the "Sâmanna phala sutta" of which Burnouf has given such interesting details in his *Lotus*, p. 449 *seq.* Fausböll also, in his notes on v. 332, rightly explains 'sâma*n*natâ' by 'srâma*n*yata.'

"Anupâdiyâno," which I have translated by "caring for nothing," has a technical meaning. It is the negative of the fourth Nidâna, the so-called Upâdâna, which Köppen has well explained by "Anhänglichkeit," taking to the world, loving the world. Köppen, *Die Religion des Buddha*, p. 610.

CHAPTER II.

ON REFLECTION.

21.

REFLECTION is the path of immortality, thoughtlessness the path of death. Those who reflect do not die, those who are thoughtless are as if dead already.

22.

Having understood this clearly, those who are advanced in reflection, delight in reflection, and rejoice in the knowledge of the Ariyas (the Elect).

(21.) "Apramâda," which Fausböll translates by *vigilantia*, Gogerly by *religion*, expresses literally the absence of that giddiness or thoughtlessness which characterizes the state of mind of worldly people. It is the first entering into one's self, and hence all virtues are said to have their root in "apramâda." (Ye keki kusalâ dhammâ sabbe te appamâdamûlakâ.) I have translated it by "reflection," sometimes by "earnestness." Immortality, amrita," is explained by Buddhaghosha as Nirvâna. "Amrita" is used, no doubt, as a synonym of Nirvâna, but this very fact shows how many conceptions entered from the very first into the Nirvâna of the Buddhists.

If it is said that those who reflect do not die, this may be understood of spiritual death. The commentator, however, takes it in a technical sense, that they are free from the two last stages of the so-called Nidânas, namely, the Garâmarana (decay and death) and the Gâti (new birth). See Köppen, *Die Religion des Buddha*, p. 609.

(22.) The Ariyas, the noble or elect, are those who have entered on the path that leads to Nirvâna. See Köppen, p. 396. Their knowledge and general status is minutely described. See Köppen, p. 436.

23.

These wise people, meditative, steady, always possessed of strong powers, attain to Nirvâna, the highest happiness.

24.

If a reflecting person has roused himself, if he is not forgetful, if his deeds are pure, if he acts with consideration, if he restrains himself, and lives according to law, — then his glory will increase.

25.

By rousing himself, by reflection, by restraint and control, the wise man may make for himself an island which no flood can overwhelm.

26.

Fools follow after vanity, men of evil wisdom. The wise man possesses reflection as his best jewel.

27.

Follow not after vanity, nor after the enjoyment of love and lust! He who reflects and meditates, obtains ample joy.

28.

When the learned man drives away vanity by reflection, he, the wise, having reached the repose of wisdom, looks down upon the fools, far from toil upon the toiling crowd, as a man who stands on a hill looks down on those who stand on the ground.

29.

Reflecting among the thoughtless, awake among the sleepers, the wise man advances like a racer leaving behind the hack.

30.

By earnestness did Maghavan (Indra) rise to the lordship of the gods. People praise earnestness; thoughtlessness is always blamed.

31.

A Bhikshu (mendicant) who delights in reflection, who looks with fear on thoughtlessness, moves about like fire, burning all his fetters, small or large.

32.

A Bhikshu (mendicant) who delights in reflection, who looks with fear on thoughtlessness, will not go to destruction — he is near to Nirvâna.

(31.) Instead of "saha*m*," which Dr. Fausböll translates by *vincens*, Dr. Weber by "conquering," I think we ought to read "*d*ahan," burning, which was evidently the reading adopted by Buddhaghosha. Mr. R. C. Childers, whom I requested to see whether the MS. at the India Office gives "saha*m*" or "daha*m*," writes that the reading "*d*aham" is as clear as possible in that MS. The fetters are meant for the senses. See *Sûtra*, 370.

CHAPTER III.

THOUGHT.

33.

As a fletcher makes straight his arrow, a wise man makes straight his trembling and unsteady thought, which is difficult to keep, difficult to turn.

34.

As a fish taken from his watery home and thrown on the dry ground, our thought trembles all over in order to escape the dominion of Mâra (the tempter).

35.

It is good to tame the mind, which is difficult to hold in and flighty, rushing wherever it listeth; a tamed mind brings happiness.

36.

Let the wise man guard his thoughts, for they are difficult to perceive, very artful, and they rush wherever they list: thoughts well guarded bring happiness.

37.

Those who bridle their mind which travels far, moves about alone, is without a body, and hides in the chamber (of the heart), will be free from the bonds of Mâra (the tempter).

(34.) On Mâra, see verses 7 and 8.

38.

If a man's thoughts are unsteady, if he does not know the true law, if his peace of mind is troubled, his knowledge will never be perfect.

39.

If a man's thoughts are not dissipated, if his mind is not perplexed, if he has ceased to think of good or evil, then there is no fear for him while he is watchful.

(39.) Fausböll traces " anavassuta," dissipated, back to the Sanskrit root " syai," to become rigid; but the participle of that root would be "sita," not " syuta." Professor Weber suggests that " anavassuta," stands for the Sanskrit " anavasruta," which he translates " unbefleckt," unspotted. If " avasruta " were the right word, it might be taken in the sense of "not fallen off, not fallen away," but it could not mean " unspotted ;" cf. " dhairya*m* no *s*usruvat," our firmness ran away. I have little doubt, however, that "avassuta" represents the Sk. " avasruta," and is derived from the root " sru" here used in its technical sense, peculiar to the Buddhist literature, and so well explained by Burnouf in his Appendix XIV. *Lotus*, p. 820. He shows that, according to Hema*k*andra and the *G*ina alankâra, âsravakshaya, Pâli âsavasa*m*khaya, is counted as the sixth abhi*gñ*â, wherever six of these intellectual powers are mentioned, instead of five. The Chinese translate the term in their own Chinese fashion by *stillationis finis*, but Burnouf claims for it the definite sense of destruction of faults or vices. He quotes from the *Lalita-vistara* (Adhyâya xxii., ed. Râjendra Lal Mittra, p. 448) the words uttered by Buddha when he arrived at his complete Buddha-hood :—

*s*ushkâ âsravâ na puna*h* sravanti

" The vices are dried up, they will not flow again," and he shows that the Pâli dictionary, the *Abhidhânappadipikâ*, explains " âsava" simply by " kâma," love, pleasure of the senses. In the Mahâparinibbâna sutta, three classes of âsava are distinguished, the kâmâsavâ, the bhavâsavâ, and the avi*gg*âsavâ. See also Burnouf, *Lotus*, p. 665.

Burnouf takes " âsrava " at once in a moral sense, but though it has that sense in the language of the Buddhists, it may have had a more material sense in the beginning. That " sru," means, to run, and is in fact a merely dialectic variety of " sru," is admitted by Burnouf. The noun " âsrava," therefore, would have meant originally, a running,

40.

Knowing that this body is (fragile) like a jar, and making this thought firm like a fortress, one should

and the question is, did it mean a running, *i. e.* a *lapsus*, or did it mean a running, *i. e.* an impetuous desire, or, lastly, did it signify originally a bodily ailment, a running sore, and assume afterwards the meaning of a moral ailment? The last view might be supported by the fact that "âsrâva" in the sense of flux or sore occurs in the *Atharvaveda*, i. 2, 4, "tad âsrâvasya bheshagam tadu rogam anînasat," this is the medicine for the sore, this destroyed the illness. But if this was the original meaning of the Buddhist "âsava," it would be difficult to explain such a word as "anâsava," faultless, nor could the participle "avasuta" or "avassuta" have taken the sense of sinful or faulty, or, at all events, engaged in worldly thoughts, attached to mundane interests. In order to get that meaning, we must assign to "âsrava" the original meaning of running towards or attending to external objects (like "sanga, âlaya," etc.) while "avasruta" would mean, carried off towards external objects, deprived of inward rest. This conception of the original purport of "â+sru" or "ava-sru" is confirmed by a statement of Colebrooke's, who, when treating of the *Gainas*, writes (*Miscellaneous Essays*, i. 382): "Âsrava is that which directs the embodied spirit (âsravayati purusham) towards external objects. It is the occupation and employment (vritti or pravritti) of the senses or organs on sensible objects. Through the means of the senses it affects the embodied spirit with the sentiment of taction, color, smell, and taste. Or it is the association or connection of body with right and wrong deeds. It comprises all the karmas, for they (âsravayanti) pervade, influence, and attend the doer, following him or attaching to him. It is a misdirection (mithyâ-pravritti) of the organs, for it is vain, a cause of disappointment, rendering the organs of sense and sensible objects subservient to fruition. Samvara is that which stops (samvrinoti) the course of the foregoing, or closes up the door or passage to it, and consists in self-command or restraint of organs internal and external, embracing all means of self-control and subjection of the senses, calming and subduing them."

For a full account of the âsravas, see also *Lalita-vistara*, ed. Calc. pp. 445 and 552, where Kshînâsrava is given as a name of Buddha.

(40.) "Anivesana" has no doubt a technical meaning, and may signify, one who has left his house, his family and friends, to become a monk. A monk shall not return to his home, but travel about; he shall be anivesana, homeless, anâgâra, houseless. But I doubt whether this can be the meaning of "anivesana" here, as the sentence, let him be an anchorite, would come in too abruptly. I translate it attack Mâra (the tempter) with the weapon of knowledge, one should watch him when conquered, and should never cease (from the fight).

41.

Before long, alas! this body will lie on the earth, despised, without understanding, like a useless log.

42.

Whatever a hater may do to a hater, or an enemy to an enemy, a wrongly-directed mind will do us greater mischief.

43.

Not a mother, not a father will do so much, nor any other relative; a well-directed mind will do us greater service.

therefore in a more general sense, let him not return or turn away from the battle, let him watch Mâra, even after he is vanquished, let him keep up a constant fight against the adversary.

CHAPTER IV.

FLOWERS.

44.

WHO shall overcome this earth, and the world of Yama (the lord of the departed), and the world of the gods? Who shall find out the plainly shown path of virtue, as a clever man finds out the (right) flower?

45.

The disciple will overcome the earth, and the world of Yama, and the world of the gods. The disciple will find out the plainly shown path of virtue, as a clever man finds out the (right) flower.

(44, 45.) If I differ from the translation of Fausböll and Weber it is because the commentary takes the two verbs, "vi*g*essati" and "pa*k*essati," to mean in the end the same thing, *i. e.*, "sa*kkh*i-karissati," he will perceive. I have not ventured to take "vi*g*essate" for "vigan-issati," but it should be remembered that the overcoming of the earth and of the worlds below and above, as here alluded to, is meant to be achieved by means of knowledge. "Pa*k*essati," he will gather (cf. vi-*k*i, *Indische Sprüche*, 4560), means also, like to gather in English, he will perceive or understand, and the "dhammapada," or path of virtue, is distinctly explained by Buddhaghosha as consisting of the thirty-seven states or stations which lead to Bodhi. See Burnouf, *Lotus*, p. 430; Hardy, *Manual*, p. 497. "Dhammapada" might, no doubt, mean also "a law-verse," but "sudesita" can hardly mean "well delivered," while, as applied to a path, it means "well pointed out" (v. 285.) Buddha himself is called "Mârga-dar*s*aka" and "Mârga-desika" (cf. *Lalita-vistara*, p. 551). Nor could one well say that a man collects one single law-verse. Hence Fausböll naturally translates *versus legis bene enarratos*, and Weber gives "Lehrsprüche" in the plural, but the origi-

46.

He who knows that this body is like froth, and has learnt that it is as unsubstantial as a mirage, will break the flower-pointed arrow of Mâra, and never see the King of Death.

47.

Death carries off a man who is gathering flowers and whose mind is distracted, as a flood carries off a sleeping village.

48.

Death subdues a man who is gathering flowers, and whose mind is distracted, before he is satiated in his pleasures.

49.

As the bee collects nectar and departs without in-

---

nal has "dhammapadam," in the sing. (47, 48). There is a curious similarity between these verses and verses 6540–41, and 9939 of the *Sânti-parva* : —

> Pushpâṇiva vikinvantam anyatragatamanasam,
> Anavâpteshu kâmeshu mrityur abhyeti mânavam.

"Death approaches man like one who is gathering flowers, and whose mind is turned elsewhere, before his desires have been fulfilled."

> Suptam vyâghram mahaugho vâ mrityur âdâya gakkhati,
> Samkinvânakam evainam kâmânâm avitriptikam.

"As a stream (carries off) a sleeping tiger, death carries off this man who is gathering flowers, and who is not satiated in his pleasures."

This last verse, particularly, seems to me clearly a translation from Pâli, and the "kam" of "samkinvânakam" looks as if put in *metri causâ*.

(46.) The flower-arrows of Mâra, the tempter, are borrowed from Kâma, the Hindu god of love. For a similar expression see *Lalitavistara*, ed. Calc. p. 40, l. 20, "mâyâmarîkisadrisâ vidyutphenopamâs kapalâh." It is on account of this parallel passage that I prefer to translate "marîki" by mirage, and not by sunbeam, as Fausböll, or by solar atom, as Weber proposes.

(48.) "Antaka," death, is given as an explanation of "Mâra" in the Amarakosha and Abhidhânappadîpika (cf. Fausböll, p. 210).

juring the flower, or its color and scent, so let the sage dwell on earth.

50.

Not the failures of others, not their sins of commission or omission, but his own misdeeds and negligences should the sage take notice of.

51.

Like a beautiful flower, full of color, but without scent, are the fine but fruitless words of him who does not act accordingly.

52.

But, like a beautiful flower, full of color and full of scent, are the fine and fruitful words of him who acts accordingly.

53.

As many kinds of wreaths can be made from a heap of flowers, so many good things may be achieved by a mortal if once he is born.

54.

The scent of flowers does not travel against the wind, nor (that of) sandal-wood, or of a bottle of Tagara oil; but the odor of good people travels even against the wind; a good man pervades every place.

55.

Sandal-wood or Tagara, a lotus flower, or a Vassikî, the scent of their excellence is peerless when their fragrance is out.

(54.) "Tagara," a plant from which a scented powder is made. "Mallaka" or "mallikâ," according to Benfey, is an oil vessel. Hence "tagaramallikâ" is probably meant for a bottle holding aromatic powder, or oil made of the Tagara.

56.

But mean is the scent that comes from Tagara and sandal-wood; the odor of excellent people rises up to the gods as the highest.

57.

Of the people who possess these excellencies, who live without thoughtlessness, and who are emancipated through true knowledge, Mâra, the tempter, never finds the way.

58, 59.

As on a heap of rubbish cast upon the highway the lily will grow full of sweet perfume and delightful, thus the disciple of the truly enlightened Buddha shines forth by his knowledge among those who are like rubbish, among the people that walk in darkness.

CHAPTER V.

THE FOOL.

60.

LONG is the night to him who is awake; long is a mile to him who is tired; long is life to the foolish who do not know the true law.

61.

If a traveller does not meet with one who is his better, or his equal, let him firmly keep to his solitary journey; there is no companionship with a fool.

62.

"These sons belong to me, and this wealth belongs to me;" with such thoughts a fool is tormented. He himself does not belong to himself; how much less sons and wealth?

63.

The fool who knows his foolishness, is wise at least so far. But a fool who thinks himself wise, he is called a fool indeed.

64.

If a fool be associated with a wise man all his life,

(60.) Life, samsâra, is the constant revolution of birth and death which goes on forever until the knowledge of the true law or the true doctrine of Buddha enables a man to free himself from samsâra, and to enter into Nirvâna. See Parable xix. p. 124.

he will perceive the truth as little as a spoon perceives the taste of soup.

65.

If an intelligent man be associated for one minute only with a wise man, he will soon perceive the truth, as the tongue perceives the taste of soup.

66.

Fools of little understanding have themselves for their greatest enemies, for they do evil deeds which must bear bitter fruits.

67.

That deed is not well done of which a man must repent, and the reward of which he receives crying and with a tearful face.

68.

No, that deed is well done of which a man does not repent, and the reward of which he receives gladly and cheerfully.

69.

As long as the evil deed done does not bear fruit, the fool thinks it is like honey; but when it ripens, then the fool suffers grief.

70.

Let a fool month after month eat his food (like an ascetic) with the tip of a blade of Kusa grass, yet is he not worth the sixteenth particle of those who have well weighed the law.

(70.) The commentator clearly takes "sa*m*khâta" in the sense of "sa*m*khyâta," not of "sa*m*skrita," for he explains it by "ñâtadhammâ tulitadhammâ." The eating with the tip of Kusa-grass has reference to the fastings performed by the Brahmans, but disapproved of, except as a moderate discipline, by the followers of Buddha. This verse seems to interrupt the continuity of the other verses which treat of the

71.

An evil deed does not turn suddenly, like milk; smouldering it follows the fool, like fire covered by ashes.

72.

And when the evil deed, after it has become known, brings sorrow to the fool, then it destroys his bright lot, nay it cleaves his head.

73.

Let the fool wish for a false reputation, for precedence among the Bhikshus, for lordship in the convents, for worship among other people!

74.

"May both the layman and he who has left the world think that this is done by me; may they be subject to me in everything which is to be done or is not to be done;" thus is the mind of the fool, and his desire and pride increase.

75.

"One is the road that leads to wealth, another the reward of evil deeds, or of the slow but sure ripening of every sinful act.

(71.) I am not at all certain of the simile, unless "mukkati," as applied to milk, can be used in the sense of changing or turning sour. In Manu, iv. 172, where a similar sentence occurs, the commentators are equally doubtful: Nâdharmas karito loke sadyah phalati gaur iva, — for an evil act committed in the world does not bear fruit at once, like a cow; or like the earth (in due season).

(72.) I take "ñattam" for "gñapitam," the causative of "gñâtam," for which in Sanskrit, too, we have the form without i, "gñaptam." This "gñaptam," made known, revealed, stands in opposition to the "khanna," covered, hid, of the preceding verse. "Sukkamsa," which Fausböll explains by "suklânsa," has probably a more technical and special meaning.

road that leads to Nirvâna;" if the Bhikshu, the disciple of Buddha, has learnt this, he will not yearn for honor, he will strive after separation from the world.

(75.) "Viveka," which in Sanskrit means chiefly understanding, has with the Buddhists the more technical meaning of separation, whether separation from the world and retirement to the solitude of the forest (kâya viveka), or separation from idle thoughts (*k*itta viveka), or the highest separation and freedom (Nirvâ*n*a).

CHAPTER VI.

THE WISE MAN.

76.

IF you see an intelligent man who tells you where true treasures are to be found, who shows what is to be avoided, and who administers reproofs, follow that wise man; it will be better, not worse, for those who follow him.

77.

Let him admonish, let him command, let him hold back from what is improper! — he will be beloved of the good, by the bad he will be hated.

78.

Do not have evil-doers for friends, do not have low people: have virtuous people for friends, have for friends the best of men.

79.

He who drinks in the Law lives happily with a serene mind: the sage rejoices always in the Law, as preached by the elect.

(78.) It is hardly possible to take "mitte kalyâne" in the technical sense of "kalyâna-mitra, "ein geistlicher Rath," a spiritual guide. Burnouf (*Introd.* p. 284) shows that in the technical sense "kalyâna-mitra" was widely spread in the Buddhist world.

(79.) The commentator clearly derives "pîti" from "pâ," to drink; if it were derived from "prî," as Professor Weber seems to

80.

Well-makers lead the water (wherever they like); fletchers bend the arrow; carpenters bend a log of wood; wise people fashion themselves.

81.

As a solid rock is not shaken by the wind, wise people falter not amidst blame and praise.

82.

Wise people, after they have listened to the laws, become serene, like a deep, smooth, and still lake.

83.

Good people walk on whatever befall, the good do not murmur, longing for pleasure; whether touched by happiness or sorrow wise people never appear elated or depressed.

---

suppose, we should expect a double p. "Ariya," elect, venerable, is explained by the commentator as referring to Buddha and other teachers.

(80.) See verse 33, and 145, the latter being a mere repetition of our verse. The "nettikâs," to judge from the commentary and from the general purport of the verse, are not simply water-carriers, but builders of canals and aqueducts, who force the water to go where it would not go by itself.

(83.) The first line is very doubtful. I have adopted, in my translation, a suggestion of Mr. Childers, who writes, "I think it will be necessary to take "sabbattha" in the sense of "everywhere," or "under every condition;" "pañkakhandâdibhedesu, sabbadhammesu," says Buddhaghosha. I do not think we need assume that B. means the word "vigahanti" to be a synonym of "vagauti." I would rather take the whole sentence together as a gloss upon the word "vagauti:" "vagantîti arahattañânena apakaddhantû khandarâgam vigahanti;" "vagauti" means that, ridding themselves of lust by the wisdom which Arhat-ship confers, they cast it away." I am inclined to think the line means "the righteous walk on (unmoved) in all the conditions of life." "Nindâ, pasamsâ, sukham, dukkham," are four of the

84.

If, whether for his own sake, or for the sake of others, a man wishes neither for a son, nor for wealth, nor for lordship, and if he does not wish for his own success by unfair means, then he is good, wise, and virtuous.

85.

Few are there among men who arrive at the other shore; the other people here run up and down the shore.

86.

But those who, when the Law has been well preached to them, follow the Law, will pass across the dominion of death, however difficult to overcome.

87, 88.

A wise man should leave the dark state (of ordinary life), and follow the bright state (of the Bhikshu). After going from his home to a homeless state, he should in his retirement look for enjoyment where there seemed to be no enjoyment. Leaving all pleasures behind, and calling nothing his own, the wise man should free himself from all the troubles of the mind.

eight lokadhammas, or earthly conditions; the remaining lokadhammas are "lâbha, alâbha, yasa, ayasa."

In v. 245, "passatâ," by a man who sees, means, by a man who sees clearly or truly. In the same manner "vrag" and "pravrag" may mean, not simply to walk, but to walk properly.

(86.) "The other shore" is meant for Nirvâna, "this shore" for common life. On reaching Nirvâna, the dominion of death is overcome. The commentator supplies "târitvâ," having crossed, in order to explain the accusative "makkudheyyam." Possibly "pâram essanti" should here be taken as one word, in the sense of overcoming.

(87, 88.) Leaving one's home is the same as joining the clergy, or becoming a mendicant, without a home or family, an "anâgâra," or anchorite. A man in that state of "viveka," or retirement (see v. 75,

89.

Those whose mind is well grounded in the elements of knowledge, who have given up all attachments, and rejoice without clinging to anything, those whose frailties have been conquered, and who are full of light, are free (even) in this world.

note), sees, that where before there seemed to be no pleasure there real pleasure is to be found, or *vice versâ*. A similar idea is expressed in verse 99. See Burnouf, *Lotus*, p. 474, where he speaks of " Le plaisir de la satisfaction, né de la distinction."

The five troubles or evils of the mind are passion, anger, ignorance, arrogance, pride. See Burnouf, *Lotus*, p. 360, and p. 443. As to " pariyodapeyya," see verse 183, and *Lotus*, pp. 523, 528; as to " akimkano," see *Mahâbh*. xii. 6568; 1240.

(89.) The elements of knowledge are the seven " Sambodhyangas," on which see Burnouf, *Lotus*, p. 796. Khînâsavâ, which I have translated by, they whose frailties have been conquered, may also be taken in a more metaphysical sense, as explained in the note to v. 39. The same applies to the other terms occurring in this verse, such as " âdâna, anupâdâya," etc. Dr. Fausböll seems inclined to take " âsava" in this passage, and in the other passages where it occurs, as the Pâli representative of " âsraya." But " âsraya," in Buddhist phraseology, means rather the five organs of sense with " manas," the soul, and these are kept distinct from the " âsavas," the inclinations, the frailties, passions, or vices. The commentary on the Abhidharma, when speaking of the Yogâkâras, says, " En réunissant ensemble les réceptacles (âsraya), les choses reçues (âsrita) et les supports (âlambana), qui sont chacun composés de six termes, on a dix-huit termes qu'on appelle 'Dhâtus' ou contenants. La collection des six réceptacles, ce sont les organes de la vue, de l'ouïe, de l'odorat, du goût, du toucher, et le " manas " (ou l'organe du cœur), qui est le dernier. La collection des six choses 'reçues, c'est la connaissance produite par la vue et par les autres sens jusqu'au 'manas' inclusivement. La collection des six supports, ce sont la forme et les autres attributs sensibles jusqu'au 'Dharma' (la loi ou l'être) inclusivement." See Burnouf, *Introduction*, p. 449.

" Parinibbuta " is again a technical term, the Sanskrit " parinivrita " meaning, freed from all worldly fetters, like " vimukta." See Burnouf, *Introduction*, p. 590.

CHAPTER VII.

THE VENERABLE.

90.

THERE is no suffering for him who has finished his journey, and abandoned grief, who has freed himself on all sides, and thrown off all fetters.

91.

They depart with their thoughts well collected, they are not happy in their abode; like swans who have left their lake, they leave their house and home.

92.

They who have no riches, who live on authorized food, who have perceived the Void, the Unconditioned, the Absolute, their way is difficult to understand, like that of birds in the ether.

(91.) "Satîmanto," Sansk. "smṛitimantaḥ," possessed of memory, but here used in the technical sense of "sati," the first of the Bodhyaṅgas. See Burnouf, *Introduction*, p. 797. Clough translates it by intense thought, and this is the original meaning of "smar," even in Sanskrit. See *Lectures on the Science of Language*, vol. ii. p. 332.

Uyyuñyanti which Buddhaghosha explains by "they exert themselves," seems to me to signify in this place "they depart," *i. e.*, they leave their family, and embrace an ascetic life. See note to verse 235.

(92.) "Suññato" (or -tâ), "animitto," and "vimokho" are three different aspects of Nirvâna. See Burnouf, *Introd.* pp. 442, 462, on sûnya. Nimitta is cause in the most general sense, what causes

93.

He whose passions are stilled, who is not absorbed in enjoyment, who has perceived the Void, the Unconditioned, the Absolute, his path is difficult to understand, like that of the birds in the ether.

94.

The gods even envy him whose senses have been subdued, like horses well broken in by the driver, who is free from pride, and free from frailty.

95.

Such a one who does his duty is tolerant like the earth, like Indra's bolt; he is like a lake without mud; no new births are in store for him.

existence to continue. The commentator explains it chiefly in a moral sense: "râgâdinimittâbhâvena animittam, tehi ka vimuttan ti animitto vimokho," *i. e.* " owing to the absence of passion and other causes, without causation; because freed from these causes, therefore it is called freedom without causation."

The simile is intended to compare the ways of those who have obtained spiritual freedom to the flight of birds, it being difficult to understand how the birds move on without putting their feet on anything. This, at least, is the explanation of the commentator. The same metaphor occurs *Mahâbh.* xii. 6763. "Gokara," which has also the meaning of food, forms a good opposition to " bhogana."

(95.) Without the hints given by the commentator, we should probably take the three similes of this verse in their natural sense, as illustrating the imperturbable state of an Arahanta, or venerable person. The earth is always represented as an emblem of patience; the bolt of Indra, if taken in its technical sense, as the bolt of a gate, might likewise suggest the idea of firmness; while the lake is a constant representative of serenity and purity. The commentator, however, suggests that what is meant is, that the earth, though flowers are cast on it, does not feel pleasure, nor the bolt of Indra displeasure, although less savory things are thrown upon it, and that in like manner a wise person is indifferent to honor or dishonor

96.

His thought is quiet, quiet are his word and deed, when he has obtained freedom by true knowledge, when he has thus become a quiet man.

97.

The man who is free from credulity, but knows the Uncreated, who has cut all ties, removed all temp-

(96.) That this very natural threefold division, thought, word, and deed, the "trividha dvâra" or the three doors of the Buddhists (Hardy, *Manual*, p. 494), was not peculiar to the Buddhists or unknown to the Brahmans, has been proved against Dr. Weber by Professor Köppen in his *Religion des Buddha*, vol. i. p. 445. He particularly called attention to Manu, xii. 4-8; and he might have added *Mahâbh.* xii. 4059, 6512, 6549, 6554; xiii. 5677, etc. Dr. Weber has himself afterwards brought forward a passage from the *Atharvaveda*, vi. 96, 3 ("yak kashushâ manasâ yak ka văkă upârima") which, however, has a different meaning. A better one was quoted by him from the Taitt, *Ar.* x. 1, 12 (yan me manasâ, vâkâ, karmanâ vâ dushkritam kritam). Similar expressions have been shown to exist in the Zendavesta, and among the Manichæans (Lassen, *Indische Alterthumskunde*, vol. iii. p. 414; see also Boehtlingk's *Dictionary*, s. v. kâya). There was no ground, therefore, for supposing that this formula had found its way into the Christian Liturgy from Persia, for as Professor Cowell remarks, Greek writers, such as Plato, employ very similar expressions, e. g. *Protag.* p. 348, 30, πρὸς ἅπαν ἔργον καὶ λόγον καὶ διανόημα. In fact, the opposition between words and deeds occurs in almost every writer, from Homer downwards; and the further distinction between thoughts and words is clearly implied in such expressions as, "they say in their heart." That the idea of sin committed by thought was not a new idea, even to the Jews, may be seen from Prov. xxiv. 9, "the thought of foolishness is sin." In the Âpastamba-sûtras, lately edited by Professor Bühler, we find the expression, "atho yatkimku manasâ vâkâ kakshushâ vâ samkalpayam dhyâyaty âhâbhi vipasyati vâ tathaiva tad bhavatîtyupadisanti;" They say that whatever a Brahman intending with his mind, voice, or eye, thinks, says, or looks, that will be. This is clearly a very different division, and it is the same which is intended in the passage from the Atharva-veda, quoted above. In the mischief done by the eye, we have the first indication of the evil eye. *Mahâbh.* xii. 3417. See *Dhammapada*, vv. 231-234.

tations, renounced all desires, he is the greatest of men.

98.

In a hamlet or in a forest, in the deep water or on the dry land, wherever venerable persons (Arahanta) dwell, that place is delightful.

99.

Forests are delightful; where the world finds no delight, there the passionless will find delight, for they look not for pleasures.

CHAPTER VIII.

THE THOUSANDS.

100.

EVEN though a speech be a thousand (of words), but made up of senseless words, one word of sense is better, which if a man hears, he becomes quiet.

101.

Even though a Gâthâ (poem) be a thousand (of words), but made up of senseless words, one word of a Gâthâ is better, which if a man hears, he becomes quiet.

102.

Though a man recite a hundred Gâthâs made up of senseless words, one word of the Law is better, which if a man hears, he becomes quiet.

103.

If one man conquer in battle a thousand times thousand men, and if another conquer himself, he is the greatest of conquerors.

104, 105.

One's own self conquered is better than all other people; not even a god, a Gandharva, not Mâra with

(100.) "Vâkâ" is to be taken as a nom. sing. fem., instead of the Sk. "vâk."

Brahman could change into defeat the victory of a man who has vanquished himself, and always lives under restraint.

106.

If a man for a hundred years sacrifice month after month with a thousand, and if he but for one moment pay homage to a man whose soul is grounded (in true knowledge), better is that homage than a sacrifice for a hundred years.

107.

If a man for a hundred years worship Agni (fire) in the forest, and if he but for one moment pay homage to a man whose soul is grounded (in true knowledge), better is that homage than sacrifice for a hundred years.

(104.) "*Gîtam*," according to the commentator, stands for *gîto* (liṅgavipallâso, *i. e.* viparyâsa); "have" is an interjection.

The Devas (gods), Gandharvas (fairies), and other fanciful beings of the Brahmanic religion, such as the Nâgas, Sarpas, Garudas, etc., were allowed to continue in the traditional language of the people who had embraced Buddhism. See the pertinent remarks of Burnouf, *Introduction*, p. 134 *seq.*, 184. On Mâra, the tempter, see v. 7. Sâstram Aiyar, *On the Gaina Religion*, p. xx., says: "Moreover as it is declared in the *Gaina* Vedas that all the gods worshipped by the various Hindu sects, namely, Siva, Brahma, Vishnu, Ganapati, Subramaniyan, and others, were devoted adherents of the abovementioned Tîrthankaras, the Gainas therefore do not consider them as unworthy of their worship; but as they are servants of Arugan, they consider them to be deities of their system, and accordingly perform certain pûgâs in honor of them, and worship them also." The case is more doubtful with orthodox Buddhists. "Orthodox Buddhists," as Mr. D'Alwis writes (*Attanagalu-vansa*, p. 55) "do not consider the worship of the Devas as being sanctioned by him who disclaimed for himself and all the Devas any power over man's soul. Yet the Buddhists are everywhere idol-worshippers. Buddhism, however, acknowledges the existence of some of the Hindu deities, and from the various friendly offices which those Devas are said to have rendered to Gotama, Buddhists evince a respect for their idols." See also *Parables*, p. 162.

108.

Whatever a man sacrifice in this world as an offering or as an oblation for a whole year in order to gain merit, the whole of it is not worth a quarter; reverence shown to the righteous is better.

109.

He who always greets and constantly reveres the aged, four things will increase to him, namely life, beauty, happiness, power.

110.

But he who lives a hundred years, vicious and unrestrained, a life of one day is better if a man is virtuous and reflecting.

111.

And he who lives a hundred years, ignorant and unrestrained, a life of one day is better, if a man is wise and reflecting.

112.

And he who lives a hundred years, idle and weak, a life of one day is better, if a man has attained firm strength.

(109.) Dr. Fausböll, in a most important note, called attention to the fact that the same verse, with slight variations, occurs in Manu. We there read, ii. 121:—

Abhivâdanasîlaya nityam vriddhopasevinah,
Katvâri sampravardhante: âyur vidyâ yaso balam.

Here the four things are, life, knowledge, glory, power.

In the Âpastamba-sûtras, 1, 2, 15, the reward promised for the same virtue is "svargam âyus ka," heaven and long life. It seems, therefore, as if the original idea of this verse came from the Brahmans, and was afterwards adopted by the Buddhists. How largely it spread is shown by Dr. Fausböll from the *Asiatic Researches*, vol. xx. p. 259, where the same verse of the Dhammapada is mentioned as being in use among the Buddhists of Siam.

(112.) On "kusito" and "hinavîriyo," see note to v. 7.

113.

And he who lives a hundred years, not seeing beginning and end, a life of one day is better if a man sees beginning and end.

114.

And he who lives a hundred years, not seeing the immortal place, a life of one day is better if a man sees the immortal place.

115.

And he who lives a hundred years, not seeing the highest law, a life of one day is better, if a man sees the highest law.

CHAPTER IX.

EVIL.

116.

IF a man would hasten towards the good, he should keep his thought away from evil; if a man does what is good slothfully, his mind delights in evil.

117.

If a man commits a sin, let him not do it again; let him not delight in sin: pain is the outcome of evil.

118.

If a man does what is good, let him do it again; let him delight in it: happiness is the outcome of good.

119.

Even an evil-doer sees happiness as long as his evil deed has not ripened; but when his evil deed has ripened, then does the evil-doer see evil.

120.

Even a good man sees evil days, as long as his good deed has not ripened; but when his good deed has ripened, then does the good man see happy days.

121.

Let no man think lightly of evil, saying in his heart, It will not come near unto me. Even by the falling of water-drops a water-pot is filled; the fool becomes full of evil, even if he gathers it little by little.

122.

Let no man think lightly of good, saying in his heart, It will not benefit me. Even by the falling of water-drops a water-pot is filled; the wise man becomes full of good, even if he gather it little by little.

123.

Let a man avoid evil deeds, as a merchant if he has few companions and carries much wealth avoids a dangerous road; as a man who loves life avoids poison.

124.

He who has no wound on his hand, may touch poison with his hand; poison does not affect one who has no wound; nor is there evil for one who does not commit evil.

125.

If a man offend a harmless, pure, and innocent person, the evil falls back upon that fool, like light dust thrown up against the wind.

126.

Some people are born again; evil-doers go to hell; righteous people go to heaven; those who are free from all worldly desires enter Nirvâna.

(124) This verse, taken in connection with what precedes, can only mean that no one suffers evil but he who has committed evil, or sin; an idea the very opposite of that pronounced in Luke xiii. 1–5.

(125.) Cf. *Indische Sprüche*, 1582; Kathâsaritsâgara, 49, 222.

(126.) For a description of hell and its long, yet not endless sufferings, see *Parables*, p. 132. The pleasures of heaven, too, are frequently described in these Parables and elsewhere. Buddha, him-

127.

Not in the sky, not in the midst of the sea, not if we enter into the clefts of the mountains, is there known a spot in the whole world where a man might be freed from an evil deed.

128.

Not in the sky, not in the midst of the sea, not if we enter into the clefts of the mountains, is there known a spot in the whole world where death could not overcome (the mortal).

---

self, enjoyed these pleasures of heaven, before he was born for the last time. It is probably when good and evil deeds are equally balanced, that men are born again as human beings; this, at least, is the opinion of the *G*ainas. Cf. *Chintâmani*, ed. H. Bower, Introd. p. xv.

CHAPTER X.

PUNISHMENT.

129.

ALL men tremble at punishment, all men fear death; remember that you are like unto them, and do not kill nor cause slaughter.

130.

All men tremble at punishment, all men love life; remember that thou art like unto them, and do not kill, nor cause slaughter.

131.

He who for his own sake punishes or kills beings longing for happiness, will not find happiness after death.

(129.) One feels tempted, no doubt, to take "upama" in the sense of the nearest (der Nächste), the neighbor, and to translate, having made oneself one's neighbor, *i. e.* "loving one's neighbor as oneself." But as "upamăm," with a short a, is the correct accusative of "upamâ," we must translate "having made oneself the likeness, the image of others," "having placed oneself in the place of others." This is an expression which occurs frequently in Sanskrit (cf. *Hitopadesa*, i. 11).

Prânâ yathâtmano sbhishtâ bhûtânâm api te tathâ,
Âtmaupamyena bhûteshu dayâm kurvanti sâdhavaḥ.

"As life is dear to oneself, it is dear also to other living beings: by comparing oneself with others, good people bestow pity on all beings."

See also *Hit.* i. 12; *Râm.* v. 23, 5, "âtmânam upamâm kritvâ sveshu dâreshu ramyatâm," "Making oneself a likeness, *i. e.*, putting oneself in the position of other people, it is right to love none but one's own wife." Dr. Fausböll has called attention to similar passages in the *Mahâbhârata*, xiii. 5569 *seq.*

(131.) Dr. Fausböll points out the striking similarity between this verse and two verses occurring in Manu and the Mahâbhârata:—

132.

He who for his own sake does not punish or kill beings longing for happiness, will find happiness after death.

133.

Do not speak harshly to anybody; those who are spoken to will answer thee in the same way. Angry speech is painful, blows for blows will touch thee.

134.

If, like a trumpet trampled under foot, thou utter not, then thou hast reached Nirvâna; anger is not known in thee.

135.

As a cowherd with his staff gathers his cows into the stable, so do Age and Death gather the life of man.

136.

A fool does not know when he commits his evil deeds: but the wicked man burns by his own deeds, as if burnt by fire.

Manu, v. 45:—
    Yo ahimsakâni bhûtâni hinasty âtmasukhekkhayâ
    Sa givams ka mritas kaiva na kvakit sukham edhate.

Mahâbh. xiii. 5568:—
    Ahimsakâni bhûtâni dandena vinihanti yah
    Âtmanah sukham ikkhan sa pretya naiva sukhî bhavet.

If it were not for "ahimsakâni," in which Manu and the Mahâbhârata agree, I should say that the verses in both were Sanskrit modifications of the Pâli original. The verse in the Mahâbhârata presupposes the verse of the Dhammapada.

(133.) See *Mahâbhârata,* xii. 4056.

(136.) The metaphor of "burning" for "suffering" is very common in Buddhist literature. Everything burns, *i. e.*, "everything suffers," was one of the first experiences of Buddha himself. See v. 146.

137.

He who inflicts pain on innocent and harmless persons, will soon come to one of these ten states:

138.

He will have cruel suffering, loss, injury of the body, heavy affliction, or loss of mind,

139.

Or a misfortune of the king, or a fearful accusation, or loss of relations, or destruction of treasures,

140.

Or lightning-fire will burn his houses; and when his body is destroyed, the fool will go to hell.

141.

Not nakedness, not platted hair, not dirt, not fasting, or lying on the earth, not rubbing with dust, not sitting motionless, can purify a mortal who has not overcome desires.

(138.) "Cruel suffering" is explained by "sîsaroga," headache, etc. "Loss" is taken for loss of money. "Injury of the body" is held to be the cutting off of the arm, and other limbs. "Heavy afflictions" are, again, various kinds of diseases.

(139.) "Misfortune of the king" may mean, a misfortune that happened to the king, defeat by an enemy, and therefore conquest of the country. "Upasarga" means accident, misfortune. Dr. Fausböll translates "râgato va upassaggam" by "fulgentis (lunæ) defectionem;" Dr. Weber, by "Bestrafung vom König." "Abbhakkhânam," Sansk. "abhyâkhyânam," is a heavy accusation for high treason, or similar offenses.

The "destruction of pleasures or treasures" is explained by gold being changed to coals (see *Parables*, p. 98), pearls to cotton-seed, corn to potsherds, and by men and cattle becoming blind, lame, etc.

(141.) Dr. Fausböll has pointed out that the same or a very similar verse occurs in a legend taken from the Divyâvadâna, and translated

142.

He who, though dressed in fine apparel, exercises tranquillity, is quiet, subdued, restrained, chaste, and has ceased to find fault with all other beings, he indeed is a Brâhma*n*a, an ascetic (*S*rama*n*a), a friar (bhikshu).

by Burnouf (*Introduction*, p. 313 *seq.*). Burnouf translates the verse: "Ce n'est ni la coutume de marcher nu, ni les cheveux nattés, ni l'usage d'argile, ni le choix des diverses espèces d'aliments, ni l'habitude de coucher sur la terre nue, ni la poussière, ni la malpropreté, ni l'attention à fuir l'abri d'un toit, qui sont capables de dissiper le trouble dans lequel nous jettent les désirs non-satisfaits; mais qu'un homme, maître de ses sens, calme, recueilli, chaste, évitant de faire du mal à aucune créature, accomplisse la Loi, et il sera, quoique paré d'ornements, un Brâhmane, un Çrama*n*a, un Religieux."

Walking naked, and the other things mentioned in our verse, are outward signs of a saintly life, and these Buddha rejects because they do not calm the passions. Nakedness he seems to have rejected on other grounds too, if we may judge from the Sumâgadhâ-avadâna: "A number of naked friars were assembled in the house of the daughter of Anâtha-pi*n*dika. She called her daughter-in-law, Sumâgadhâ, and said, 'Go and see those highly respectable persons.' Sumâgadhâ, expecting to see some of the saints, like *S*âriputra, Maudgalyâyana, and others, ran out full of joy. But when she saw these friars with their hair like pigeon wings, covered by nothing but dirt, offensive, and looking like demons, she became sad. 'Why are you sad?' said her mother-in-law. Sumâgadhâ replied, 'O, mother, if these are saints, what must sinners be like?'"

Burnouf (*Introduction*, p. 312) supposed that the *G*ainas only, and not the Buddhists, allowed nakedness. But the *G*ainas, too, do not allow it universally. They are divided into two parties, the *S*vetambaras and Digambaras. The *S*vetambaras, clad in white, are the followers of Parsvanâtha, and wear clothes. The Digambaras, *i. e.* sky-clad, disrobed, are followers of Mahâvîra, and resident chiefly in Southern India. At present they, too, wear clothing, but not when eating. See *Sâstram Aiyar*, p. xxi.

The "*ga*tà," or the hair platted and gathered up in a knot, was a sign of a *S*aiva ascetic. The sitting motionless is one of the postures assumed by ascetics. Clough explains "ukku*t*ika" as the act of sitting on the heels; Wilson gives for "utka*t*ukâsana," "sitting on the hams." See Fausböll, note on verse 140.

(142.) As to "da*n*danidhâna," see *Mahâbh.* xii. 6559.

143.

Is there in this world any man so restrained by humility that he does not mind reproof, as a well-trained horse the whip ?

144.

Like a well-trained horse when touched by the whip, be ye active and lively, and by faith, by virtue, by energy, by meditation, by discernment of the law you will overcome this great pain (of reproof), perfect in knowledge and in behavior, and never forgetful.

145.

Well-makers lead the water (wherever they like), fletchers bend the arrow ; carpenters break a log of wood ; wise people fashion themselves.

(143, 144.) I am very doubtful as to the real meaning of these verses. I think their object is to show how reproof or punishment should be borne. I therefore take " bhadra assa " in the sense of a well-broken or well-trained, not in the sense of a spirited horse. "Hrî," no doubt, means generally "shame," but it also means "humility," or "modesty." However, I give my translation as conjectural only, for there are several passages in the commentary which I do not understand.

(145.) The same as verse 80.

CHAPTER XI.

OLD AGE.

146.

How is there laughter, how is there joy, as this world is always burning? Why do you not seek a light, ye who are surrounded by darkness?

147.

Look at this dressed-up lump, covered with wounds, joined together, sickly, full of many thoughts, which has no strength, no hold!

148.

This body is wasted, full of sickness, and frail; this heap of corruption breaks to pieces, the life in it is death.

149.

Those white bones, like gourds thrown away in the autumn, what pleasure is there in looking at them!

150.

After a frame has been made of the bones, it is covered with flesh and blood, and there dwell in it old age and death, pride and deceit.

(146.) Dr. Fausböll translates " semper exardescit recordatio ; " Dr. Weber, " da's doch beständig Kummer giebt." The commentator explains, " as this abode is always lighted by passion and the other fires." Cf. Hardy, *Manual*, p. 495.

(150.) The expression " ma*m*salohitalepanam " is curiously like the

151.

The brilliant chariots of kings are destroyed, the body also approaches destruction, but the virtues of good people never approach destruction, thus do the good say to the good.

152.

A man who has learnt little, grows old like an ox; his flesh grows, but his knowledge does not grow.

153, 154.

Without ceasing shall I run through a course of many births, looking for the maker of this tabernacle,—and painful is birth again and again. But now, maker of the tabernacle, thou hast been seen; thou shalt not make up this tabernacle again. All thy rafters are broken, thy ridge-pole is sundered; the mind, being sundered, has attained to the extinction of all desires.

expression used in Manu, vi. 76, "mâmsasonitalpanam," and in several passages of the *Mahâbhârata*, xii. 12053, 12462, as pointed out by Dr. Fausböll.

(153, 154.) These two verses are famous among Buddhists, for they are the words which the founder of Buddhism is supposed to have uttered at the moment he attained to Buddhahood. See Spence Hardy, *Manual*, p. 180. According to the *Lalita-vistara*, the words uttered on that solemn occasion were those quoted in the note to verse 39. Though the purport of both is the same, the tradition preserved by the Southern Buddhists shows greater vigor than that of the North.

"The maker of the tabernacle" is explained as a poetical expression for the cause of new births, at least according to the views of Buddha's followers, whatever his own views may have been. Buddha had conquered Mâra, the representative of worldly temptations, the father of worldly desires, and as desires (tanhâ) are, by means of "upâdâna" and "bhava," the cause of "gâti," or birth, the destruction of desires and the defeat of Mâra are really the same thing, though expressed differently in the philosophical and legendary

155.

Men who have not observed proper discipline, and have not gained wealth in their youth, they perish like old herons in a lake without fish.

156.

Men who have not observed proper discipline, and have not gained wealth in their youth; they lie like broken bows, sighing after the past.

language of the Buddhists. Ta*n*hâ, thirst or desire, is mentioned as serving in the army of Mâra. *Lotus*, p. 443. There are some valuable remarks of Mr. D'Alwis on these verses in the *Attanugaluvansa*, p. cxxviii. This learned scholar points out a certain similarity in the metaphors used by Buddha, and some verses in Manu, vi. 76, 77. See also *Mahâbh.* xii. 12463-4. Mr. D'Alwis' quotation, however, from Pâ*n*ini, iii. 2, 112, proves in no way that "sandhavissan," or any other future can, if standing by itself, be used in a past sense. Pâ*n*ini speaks of "bhûta*a*nadyatana," and he restricts the use of the future in a past sense to cases where the future follows verbs expressive of recollection, etc.

(155.) On "*gh*âyanti," *i. e.* "kshâyanti," see Dr. Bollensen's learned remarks, *Zeitschrift der Deutschen Morgenl. Gesellschaft*, xviii. 834, and Boehtlingk-Roth, *s. v.* "kshâ."

CHAPTER XII.

SELF.

157.

IF a man hold himself dear, let him watch himself carefully; during one at least out of the three watches a wise man should be watchful.

158.

Let each man first direct himself to what is proper, then let him teach others; thus a wise man will not suffer.

159.

Let each man make himself as he teaches others to be; he who is well subdued may subdue (others); one's own self is difficult to subdue.

160.

Self is the lord of self, who else could be the lord? With self well-subdued, a man finds a lord such as few can find.

161.

The evil done by oneself, self-begotten, self-bred, crushes the wicked, as a diamond breaks a precious stone.

(157.) The three watches of the night are meant for the three stages of life.

162.

He whose wickedness is very great brings himself down to that state where his enemy wishes him to be, as a creeper does with the tree which it surrounds.

163.

Bad deeds, and deeds hurtful to ourselves, are easy to do; what is beneficial and good, that is very difficult to do.

164.

The wicked man who scorns the rule of the venerable (Arahat), of the elect (Ariya), of the virtuous, and follows false doctrine, he bears fruit to his own destruction, like the fruits of the Ka*tth*aka reed.

165.

By oneself the evil is done, by oneself one suffers; by oneself evil is left undone, by oneself one is purified. Purity and impurity belong to oneself, no one can purify another.

166.

Let no one forget his own duty for the sake of

(164.) The reed either dies after it has borne fruit or is cut down for the sake of its fruit.

"Di*tth*i," literally view, is used even by itself, like the Greek "hairesis" in the sense of heresy (see Burnouf, *Lotus*, p. 444). In other places a distinction is made between "mi*khh*âdi*tth*i" (vv. 167, 316) and "sammâdi*tth*i" (v. 319). If "arahata*m* ariyàna*m*" are used in their technical sense, we should translate "the reverend Arhats,"—"Arhat" being the highest degree of the four orders of Ariyas, namely, Srotaâpanna, Sa*k*ridâgâmin, Anâgâmin, and Arhat. See note to v. 178.

(166.) "Attha," lit. "object," must be taken in a moral sense, as "duty" rather than as "advantage." The story which Buddhaghosha tells of the "Thera Attadattha" gives a clew to the origin another's, however great; let a man, after he has discerned his own duty, be always attentive to his duty.

of some of his parables, which seem to have been invented to suit the text of the Dhammapada rather than *vice versâ*. A similar case occurs in the commentary to verse 227.

CHAPTER XIII.

THE WORLD.

167.

Do not follow the evil law! Do not live on in thoughtlessness! Do not follow false doctrine! Be not a friend of the world.

168.

Rouse thyself! do not be idle! Follow the law of virtue! The virtuous lives happily in this world and in the next.

169.

Follow the law of virtue; do not follow that of sin. The virtuous lives happily in this world and in the next.

170.

Look upon the world as a bubble, look upon it as a mirage: the king of death does not see him who thus looks down upon the world.

171.

Come, look at this glittering world, like unto a royal chariot; the foolish are immersed in it, but the wise do not cling to it.

172.

He who formerly was reckless and afterwards became sober, brightens up this world, like the moon when freed from clouds.

173.

He whose evil deeds are covered by good deeds, brightens up this world, like the moon when freed from clouds.

174.

This world is dark, few only can see here; a few only go to heaven, like birds escaped from the net.

175.

The swans go on the path of the sun, they go through the ether by means of their miraculous power; the wise are led out of this world, when they have conquered Mâra and his train.

176.

If a man has transgressed one law, and speaks lies, and scoffs at another world, there is no evil he will not do.

177.

The uncharitable do not go to the world of the gods; fools only do not praise liberality; a wise man rejoices in liberality, and through it becomes blessed in the other world.

178.

Better than sovereignty over the earth, better than going to heaven, better than lordship over all worlds, is the reward of the first step in holiness.

(175.) "Hamsa" may be meant for the bird, whether flamingo, or swan, or ibis (see Hardy, *Manual*, p. 17), but it may also, I believe, be taken in the sense of saint. As to "iddhi," magical power, *i. e.* "riddhi," see Burnouf, *Lotus*, p. 310; Spence Hardy, *Manual*, pp. 498 and 504; *Legends*, pp. 55, 177. See note to verse 254.

(178.) "Sotâpatti," the technical term for the first step in the path that leads to Nirvâna. There are four such steps, or stages, and on entering each, a man receives a new title: —

1. The "Srota âpanna," lit. he who has got into the stream. A man may have seven more births before he reaches the other shore, *i. e.*, "Nirvâna."

2. "Sakridâgâmin," lit. he who comes back once, so called because, after having entered this stage, a man is born only once more among men or gods.

3. "Anâgâmin," lit. he who does not come back, so called because, after this stage, a man cannot be born again in a lower world, but can only enter a Brahman world before he reaches Nirvâna.

4. "Arhat," the venerable, the perfect, who has reached the highest stage that can be reached, and from which Nirvâna is perceived (sukkhavipassanâ, *Lotus*, p. 849). See Hardy, "*Eastern Monachism*, p. 280; Burnouf, *Introduction*, p. 209; Köppen, p 398; D'Alwis, *Attanugaluvansa*, p. cxxiv.

CHAPTER XIV.

THE AWAKENED (BUDDHA).

179.

HE whose conquest is not conquered again, whose conquest no one in this world escapes, by what path can you lead him, the Awakened, the Omniscient, into a wrong path?

180.

He whom no desire with its snares and poisons can lead astray, by what path can you lead him, the Awakened, the Omniscient, into a wrong path?

181.

Even the gods envy those who are awakened and not forgetful, who are given to meditation, who are wise, and who delight in the repose of retirement (from the world).

(179, 180.) These two verses, though their general meaning seems clear, contain many difficulties which I do not at all pretend to solve. "Buddha," the Awakened, is to be taken as an appellative rather than as the proper name of the "Buddha." It means, anybody who has arrived at complete knowledge. "Anantago*k*aram" I take in the sense of, possessed of unlimited knowledge. "Apadam," which Dr. Fausböll takes as an epithet of Buddha and translates by *non investigabilis*, I take as an accusative governed by "nessatha," and in the sense of wrong place (uppatha, v. 309, p. 396, l. 2) or sin.

The second line of verse 179 is most difficult. The commentator seems to take it in the sense of "in whose conquest nothing is wanting," "who has conquered all sins and all passions." In that case we should have to supply "kileso" (masc.) or "râgo," or take "ko*k*i" in the sense of any enemy. Cf. v. 105.

182.

Hard is the conception of men, hard is the life of mortals, hard is the hearing of the True Law, hard is the birth of the Awakened (the attainment of Buddhahood).

183.

Not to commit any sin, to do good, and to purify one's mind, that is the teaching of the Awakened.

184.

The Awakened call patience the highest penance, long-suffering the highest Nirvâna; for he is not an anchorite (Pravragita) who strikes others, he is not an ascetic (Sramana) who insults others.

185.

Not to blame, not to strike, to live restrained under the law, to be moderate in eating, to sleep and eat

---

(183.) This verse is again one of the most solemn verses among the Buddhists. According to Csoma de Körös, it ought to follow the famous Âryâ stanza, "Ye dhammâ" (*Lotus*, p. 522), and serve as its complement. But though this may be the case in Tibet, it was not so originally. Burnouf has fully discussed the metre and meaning of our verse on pp. 527, 528 of his *Lotus*. He prefers "sakittaparidamanam," which Csoma translated by "the mind must be brought under entire subjection" (svakittaparidamanam), and the late Dr. Mill by "proprii intellectus subjugatio." But his own MS. of the "Mahâpadhâna sutta" gave likewise "sakittapariyodapanam," and this is no doubt the correct reading. See D'Alwis, *Attanugaluvansa*, p. cxxix. We found "pariyodappeya" in verse 88, in the sense of freeing oneself from the troubles of thought. The only question is whether the root "dâ," with the prepositions "pari" and "ava," should be taken in the sense of cleansing oneself from, or cutting oneself out from. I prefer the former conception, the same which in Buddhist literature has given rise to the name Avadâna, a legend, originally a pure and virtuous act, and ἀρίστεια, afterwards a sacred story, and possibly a story the hearing of which purifies the mind. See Boehtlingk-Roth, s. v. "avadâna."

alone, and to dwell on the highest thoughts, — this is the teaching of the Awakened.

186.

There is no satisfying lusts, even by a shower of gold pieces; he who knows that lusts have a short taste and cause pain, he is wise.

187.

Even in heavenly pleasures he finds no satisfaction, the disciple who is fully awakened delights only in the destruction of all desires.

188.

Men, driven by fear, go to many a refuge, to mountains and forests, to groves and sacred trees.

189.

But that is not a safe refuge, that is not the best refuge; a man is not delivered from all pains after having gone to that refuge.

(185.) "Pâtimokkhe," under the law, *i. e.*, according to the law, the law which leads to "Moksha," or freedom. "Prâtimoksha" is the title of the oldest collection of the moral laws of the Buddhists (Burnouf, *Introduction*, p. 300; Bigandet, *The Life of Guadama*, p. 439), and as it was common both to the Southern and the Northern Buddhists, "pâtimokkhe" in our passage may possibly be meant, as Professor Weber suggests, as the title of that very collection. The commentator explains it by "*getthaksîla*" and "pâtimokkhasîla." I take "sayanâsam" for "sayanâsanam;" see *Mahâb.* xii. 6684. In xii. 9978, however, we find also "sayyâsane."

(187.) There is a curious similarity between this verse and verse 6503 (9919) of the *Sântiparva*: —

        Ya*k* k*a* kâmasukha*m* loke, ya*k* k*a* divyam mahat sukham,
        Trishnâkshayasukhasyaite nârhata*h* shodasim kalâm.

"And whatever delight of love there is on earth, and whatever is the great delight in heaven, they are not worth the sixteenth part of the pleasure which springs from the destruction of all desires."

190.

He who takes refuge with Buddha, the Law, and the Church; he who, with clear understanding, sees the four holy truths: —

191.

Namely, pain, the origin of pain, the destruction of pain, and the eightfold holy way that leads to the quieting of pain, —

192.

That is the safe refuge, that is the best refuge; having gone to that refuge, a man is delivered from all pain.

193.

A supernatural person is not easily found, he is not born everywhere. Wherever such a sage is born, that race prospers.

194.

Happy is the arising of the Awakened, happy is the teaching of the True Law, happy is peace in the church, happy is the devotion of those who are at peace.

(188–192.) These verses occur in Sanskrit in the "Prâtihârya-sûtra," translated by Burnouf, *Introduction*, pp. 162–189; see p. 186. Burnouf translates "rukkhaketyâni" by "arbres consacrés;" properly, sacred shrines under or near a tree.

(190.) Buddha, Dharma, and Sangha are called the "Trisarana" (cf. Burnouf, *Introd.* p. 630). The four holy truths are the four statements that there is pain in this world, that the source of pain is desire, that desire can be annihilated, that there is a way (shown by Buddha) by which the annihilation of all desires can be achieved, and freedom be obtained. That way consists of eight parts. See Burnouf, *Introduction*, p. 630. The eightfold way forms the subject of chapter xviii. See also *Chips from a German Workshop*, 2d. ed. vol. i. p. 251 *seq.*

195, 196.

He who pays homage to those who deserve homage, whether the awakened (Buddha) or their disciples, those who have overcome the host (of evils), and crossed the flood of sorrow, he who pays homage to such as have found deliverance and know no fear, his merit can never be measured by anybody.

CHAPTER XV.

HAPPINESS.

197.

LET us live happily, then, not hating those who hate us! let us dwell free from hatred among men who hate!

198.

Let us live happily, then, free from ailments among the ailing! let us dwell free from ailments among men who are ailing!

199.

Let us live happily, then, free from greed among the greedy! let us dwell free from greed among men who are greedy!

200.

Let us live happily, then, though we call nothing our own! We shall be like the bright gods, feeding on happiness!

(198.) The ailment here meant is moral rather than physical. Cf. *Mahâbh.* xii. 9924, "samprasânto nirâmayah;" 9925, "yo sauprânântiko rogas tâm trishnâm tyagatah sukham."

(200.) The words placed in the mouth of the king of Videha, while his residence Mithilâ was in flames, are curiously like our verse; cf. *Mahâbh.* xii. 9917: —

> Susukham vata gîvâmi yasya me nâsti kimkana.
> Mithilâyâm pradîptâyâm na me dahyati kimkana.

"I live happily, indeed, for I have nothing; while Mithilâ is in flames, nothing of mine is burning."

The "âbhassara," *i. e.* "âbhâsvara," the bright gods, are frequently mentioned. Cf. Burnouf, *Introd.* p. 611.

201.

Victory breeds hatred, for the conquered is unhappy. He who has given up both victory and defeat, he, the contented, is happy.

202.

There is no fire like passion: there is no unlucky die like hatred; there is no pain like this body; there is no happiness like rest.

203.

Hunger is the worst of diseases, the body the greatest of pains; if one knows this truly, that is Nirvâna, the highest happiness.

(202.) I take "kali" in the sense of an unlucky die which makes a player lose his game. A real simile seems wanted here, as in v. 252, where, for the same reason, I translate "graha" by "shark," not by "captivitas," as Dr. Fausböll proposes. The same scholar translates "kali" in our verse by "peccatum." If there is any objection to translating "kali" in Pâli by unlucky die, I should still prefer to take it in the sense of the age of depravity, or the demon of depravity.

"Body" for "khandha" is a free translation, but it is difficult to find any other rendering. According to the Buddhists each sentient being consists of five "khandha" (skandha), or branches, the organized body (rûpa khandha) with its four internal capacities of sensation (vedanâ), perception (samgñâ), conception (samskâra), knowledge (vigñâna). See Burnouf, *Introd.* pp. 589, 634; *Lotus,* p. 335.

(203.) It is difficult to give an exact rendering of "samskâra," which I have translated sometimes by "body" or "created things," sometimes by "natural desires." "Samskâra" is the fourth of the five "khandhas," but the commentator takes it here, as well as in v. 255, for the five "khandhas" together, in which case we can only translate it by body, or created things. There is, however, another "samskâra," that which follows immediately upon "avidyâ," ignorance, as the second of the "nidânas," or causes of existence, and this too might be called the greatest pain, considering that it is the cause of birth, which is the cause of all pain. Burnouf, *Lotus*, pp. 109, 827

204.

Health is the greatest of gifts, contentedness the best riches; trust is the best of relatives, Nirvâ*n*a, the highest happiness.

205.

He who has tasted the sweetness of solitude and tranquillity, is free from fear and free from sin, while he tastes the sweetness of drinking in the Law.

206.

The sight of the elect (Arya) is good, to live with them is always happiness; if a man does not see fools, he will be truly happy.

207.

He who walks in the company of fools suffers a long way; company with fools, as with an enemy, is always

---

says," L'homme des Buddhistes qui, doué intérieurement de l'idée de la forme, voit au dehors des formes, et, après les avoir vaincues, se dit: je connais, je vois, ressemble singulièrement au 'sujet victorieux de chaque objectivité qui demeure le sujet triomphant de toutes choses.'"

'Samskâra' seems sometimes to have a different and less technical meaning, and to be used in the sense of conceptions, plans, desires, as, for instance, in v. 368, where "sa*m*khârâ*n*am khayam" is used much like "ta*m*hâkhaya." Desires, however, are the result of "sa*m*khâra," and if the sa*m*khâras are destroyed, desires cease; see v. 154, "visa*m*khâragata*m* *k*itta*m* ta*m*hâna*m* khayam a*ggh*agâ." Again, in his comment on v. 75, Buddhaghosha says, "upadhiviveko sa*m*khârasam*g*a*n*ikam vinodeti;" and again, "upadhiviveko *k*a nirupadhînâ*m* pug*g*alâna*m* visa*m*khâragatanâm."

For a similar sentiment, see Stanislas Julien, *Les Avadanas*, vol. i. p. 40, "Le corps est la plus grande source de souffrance," etc. I should say that "khandha" in v. 202, and "sa*m*khâra" in v. 203, are nearly, if not quite, synonymous. I should prefer to read "*g*iga*kkh*âparamâ" as a compound. "*G*iga*kkh*â," or as it is written in one MS., "diga*kkh*â," Sk. "*g*ighatsâ" means not only hunger, but appetite, desire.

painful; company with the wise is pleasure, like meeting with kinsfolk.

208.

Therefore, one ought to follow the wise, the intelligent, the learned, the much enduring, the dutiful, the elect; one ought to follow a good and wise man, as the moon follows the path of the stars.

(208.) I should like to read "sukho *ka* dhîrasa*m*vâso."

CHAPTER XVI.

PLEASURE.

209.

HE who gives himself to vanity, and does not give himself to meditation, forgetting the real aim (of life) and grasping at pleasure, will in time envy him who has exerted himself in meditation.

210.

Let no man ever look for what is pleasant, or what is unpleasant. Not to see what is pleasant is pain, and it is pain to see what is unpleasant.

211.

Let, therefore, no man love anything; loss of the beloved is evil. Those who love nothing, and hate nothing, have no fetters.

212.

From pleasure comes grief, from pleasure comes fear; he who is free from pleasure knows neither grief nor fear.

213.

From affection comes grief, from affection comes fear; he who is free from affection knows neither grief nor fear.

214.

From lust comes grief, from lust comes fear; he who is free from lust knows neither grief nor fear.

215.

From love comes grief, from love comes fear; he who is free from love knows neither grief nor fear.

216.

From greed comes grief, from greed comes fear; he who is free from greed knows neither grief nor fear.

217.

He who possesses virtue and intelligence, who is just, speaks the truth, and does what is his own business, him the world will hold dear.

218.

He in whom a desire for the Ineffable (Nirvâna) has sprung up, who is satisfied in his mind, and whose thoughts are not bewildered by love, he is called Ûrdhvamsrotas (carried upwards by the stream).

(218.) "Ûrdhvamsrotas," or "uddhamsoto," is the technical name for one who has reached the world of the "Avrihas" (Aviha), and is proceeding to that of the "Akanishthas" (Akanittha). This is the last stage before he reaches the formless world, the "Arûpadhâtu." See *Parables*, p. 123; Burnouf, *Introd.* p. 599. Originally "ûrdhvamsrotas" may have been used in a less technical sense, meaning one who swims against the stream, and is not carried away by the vulgar passions of the world.

219.

Kinsfolk, friends, and lovers salute a man who has been long away, and returns safe from afar.

220.

In like manner his good works receive him who has done good, and has gone from this world to the other; as kinsmen receive a friend on his return.

CHAPTER XVII.

ANGER.

221.

LET a man leave anger, let him forsake pride, let him overcome all bondage! No sufferings befall the man who is not attached to either body or soul, and who calls nothing his own.

222.

He who holds back rising anger like a rolling chariot, him I call a real driver; other people are but holding the reins.

223.

Let a man overcome anger by love, let him overcome evil by good; let him overcome the greedy by liberality, the liar by truth!

224.

Speak the truth, do not yield to anger; give, if thou art asked, from the little thou hast; by those steps thou wilt go near the gods.

(221.) "Body and soul" is the translation of "nâma-rûpa," lit. "name and form," the ninth of the Buddhist Nidânas. Cf. Burnouf, *Introd.* p. 501; see also Gogerly, *Lecture on Buddhism*, and Bigant, *The Life of Gaudama*, p. 454.

(223.) *Mahâbh.* xii. 3550, "asâdhum sadhunâ gayet."

225.

The sages who injure nobody, and who always control their body, they will go to the unchangeable place (Nirvâna), where if they have gone, they will suffer no more.

226.

Those who are always watchful, who study day and night, and who strive after Nirvâna, their passions will come to an end.

227.

This is an old saying, O Atula, this is not only of to-day: "They blame him who sits silent, they blame him who speaks much, they also blame him who says little; there is no one on earth who is not blamed."

228.

There never was, there never will be, nor is there now, a man who is always blamed, or a man who is always praised.

229, 230.

But he whom those who discriminate praise continually day after day, as without blemish, wise, rich in knowledge and virtue, who would dare to blame him, like a coin made of gold from the Gambû river?

(227.) It appears from the commentary that "porânam" and "aggatanam" are neuters, referring to what happened formerly and what happens to-day, and that they are not to be taken as adjectives referring to "âsînam," etc. The commentator must have read "atula" instead of "atulam," and he explains it as the name of a pupil whom Gautama addressed by that name. This may be so (see note to verse 166); but "atula" may also be taken in the sense of incomparable (Mahâbh. xiii. 1937), and in that case we ought to supply, with Professor Weber, some such word as "saw" or "saying."

(230.) The Brahman worlds are higher than the Deva worlds as

Even the gods praise him, he is praised even by Brahman.

231.

Beware of bodily anger, and control thy body! Leave the sins of the body, and with thy body practice virtue!

232.

Beware of the anger of the tongue, and control thy tongue! Leave the sins of the tongue, and practice virtue with thy tongue!

233.

Beware of the anger of the mind, and control thy mind! Leave the sins of the mind, and practice virtue with thy mind!

234.

The wise who control their body, who control their tongue, the wise who control their mind, are indeed well controlled.

---

the Brahman is higher than a Deva. See Hardy, *Manual*, p. 25; Burnouf, *Introduction*, pp. 134, 184.

CHAPTER XVIII.

IMPURITY.

235.

THOU art now like a sear leaf, the messengers of Death (Yama) have come near to thee; thou standest at the door of thy departure, and thou hast no provision for thy journey.

236.

Make thyself an island, work hard, be wise! When thy impurities are blown away, and thou art free from guilt, thou wilt enter into the heavenly world of the Elect (Ariya).

237.

Thy life has come to an end, thou art come near to Death (Yama), there is no resting-place for thee on the road, and thou hast no provision for thy journey.

238.

Make thyself an island, work hard, be wise! When thy impurities are blown away, and thou art free from guilt, thou wilt not enter again into birth and decay.

(235.) "Uyyoga" seems to means "departure." See Buddhaghosha's commentary on verse 152, p. 319, l. 1; Fausböll, *Five Gâtakas*, p. 35.

(236) An "island," for a drowning man to save himself. See verse 25. "Dipamkara" is the name of one of the former Buddhas, and it is also used as an appellative of the Buddha.

239.

Let a wise man blow off the impurities of his soul, as a smith blows off the impurities of silver, one by one, little by little, and from time to time.

240.

Impurity arises from the iron, and, having arisen from it, it destroys it; thus do a transgressor's own works lead him to the evil path.

241.

The taint of prayers is non-repetition; the taint of houses, non-repair; the taint of the body is sloth, the taint of a watchman thoughtlessness.

242.

Bad conduct is the taint of woman, greediness the taint of a benefactor; tainted are all evil ways, in this world and in the next.

243.

But there is a taint worse than all taints, ignorance is the greatest taint. O mendicants! throw off that taint, and become taintless!

244.

Life is easy to live for a man who is without shame, a crow hero, a mischief-maker, an insulting, bold, and wretched fellow.

(244.) "Pakkhandin" is identified by Dr. Fausböll with "praskandin," one who jumps forward, insults, or, as Buddhaghosha explains it, one who meddles with other people's business, an interloper. At all events, it is a term of reproach, and, as it would seem, of theological reproach.

245.

But life is hard to live for a modest man, who always looks for what is pure, who is disinterested, quiet, spotless, and intelligent.

246.

He who destroys life, who speaks untruth, who takes in this world what is not given him, who takes another man's wife;

247.

And the man who gives himself to drinking intoxicating liquors, he, even in this world, digs up his own root.

248.

O man, know this, that the unrestrained are in a bad state; take care that greediness and vice do not bring thee to grief, for a long time!

249.

The world gives according to their faith or according to their pleasure: if a man frets about the food and the drink given to others, he will find no rest either by day or by night.

250.

He in whom that feeling is destroyed, and taken out with the very root, finds rest by day and by night.

(246.) On the five principal commandments which are recapitulated in verses 246 and 247, see *Parables*, p. 153.

(248.) Cf. *Mahâbhârata*, xii. 4055, "yeshâm vrittis kâ samyatâ." See also v. 307.

(249.) This verse has evidently regard to the feelings of the Bhikshus or mendicants who receive either much or little, and who are exhorted not to be envious if others receive more than they themselves. Several of the parables illustrate this feeling.

251.

There is no fire like passion, there is no shark like hatred, there is no snare like folly, there is no torrent like greed.

252.

The fault of others is easily perceived, but that of oneself is difficult to perceive; the faults of others one lays open as much as possible, but one's own fault one hides as a cheat hides the bad die from the gambler.

253.

If a man looks after the faults of others, and is always inclined to detract, his own weaknesses will grow, and he is far from the destruction of weakness.

254.

There is no path through the air, a man is not a Sramana by outward acts. The world delights in vanity, the Tathâgatas (the Buddhas) are free from vanity.

(251.) Dr. Fausböll translates "gaho" by "captivitas," Dr. Weber by "fetter." I take it in the same sense as "grâha" in Manu, vi. 78; and Buddhaghosha does the same, though he assigns to "grâha" a more general meaning, namely, anything that seizes, whether an evil spirit (yakkha), a serpent (agagara) or a crocodile (kumbhîla).

Greed or thirst is represented as a river in "Lalita-vistara," ed. Calc. p. 482, "trishnâ-nadî tivegâ prasoshitâ me gñânasûryena," the wild river of thirst is dried up by the sun of my knowledge.

(253) As to "âsava," "weakness," see note to v. 39.

(254.) I have translated this verse very freely, and not in accordance with Buddhaghosha's commentary. Dr. Fausböll proposed to translate: "No one who is outside the Buddhist community can walk through the air, but only a Sramana;" and the same view is taken by Professor Weber, though he arrives at it by a different construction. Now it is perfectly true that the idea of magical powers (riddhi) which enable saints to walk through the air, etc, occurs in the Dhammapada, see v. 175, note. But the Dhammapada may contain earlier

255.

There is no path through the air, a man is not a *S*ramana by outward acts. No creatures are eternal; but the Awakened (Buddha) are never shaken.

and later verses, and in that case our verse might be an early protest on the part of Buddha against the belief in such miraculous powers. We know how Buddha himself protested against his disciples being called upon to perform vulgar miracles. "I command my disciples not to work miracles," he said, "but to hide their good deeds, and to show their sins." Burnouf, *Introduction*, p. 170. It would be in harmony with this sentiment if we translated our verse as I have done. As to "bahira," I should take it in the sense of "external," as opposed to "adhyâtmika," or "internal;" and the meaning would be, a "*S*ramana is not a *S*ramana by outward acts, but by his heart."

"Prapa*n*ka," which I have here translated by "vanity, seems to include the whole host of human weaknesses; cf. v. 196, where it is explained by "ta*m*hâdi*tth*imânapapa*n*ka;" in our verse by "ta*m*hâdisu papa*n*kesu." Cf. *Lalita-vistara*, p. 564, "anâlaya*m* nishprapa*n*k'am anutpâdam asambhavam (dharmak'akram)." As to "Tathogata," a name of Buddha, cf. Burnouf, *Introd.* p. 75.

(259.) "Sa*m*khârâ" for "sa*m*skârâ; cf. note to v. 203

CHAPTER XIX.

THE JUST.

256, 257.

A MAN is not a just judge if he carries a matter by violence; no, he who distinguishes both right and wrong, who is learned and leads others, not by violence, but by law and equity, he who is a guardian of the law and intelligent, he is called Just.

258.

A man is not learned because he talks much; he who is patient, free from hatred and fear, he is called learned.

259.

A man is not a supporter of the law because he talks much; even if a man has learned little, but sees the law bodily, he is a supporter of the law, a man who never neglects the law.

260.

A man is not an elder because his head is gray; his age may be ripe, but he is called "Old-in-vain."

261.

He in whom there is truth, virtue, love, restraint,

(259.) Buddhaghosha here takes law (dhamma) in the sense of the four great truths, see note to v. 190. Could "dhammam kâyena passati" mean, he observes the law in his acts? Hardly, if we compare expressions like "dhammam vipassato," v. 373.

moderation, he who is free from impurity and is wise, he is called an " Elder."

262.

An envious, greedy, dishonest man does not become respectable by means of much talking only, or by the beauty of his complexion.

263.

He in whom all this is destroyed, taken out with the very root, he, freed from hatred and wise, is called " Respectable."

264.

Not by tonsure does an undisciplined man who speaks falsehood, become a *S*ramana; can a man be a *S*ramana who is still held captive by desire and greediness?

265.

He who always quiets the evil, whether small or large, he is called a *S*ramana (a quiet man), because he has quieted all evil.

(265.) This is curious etymology, because it shows that at the time when this verse was written, the original meaning of " sramana " had been forgotten. " *S*rama*n*a " meant originally, in the language of the Brahmans, a man who performed hard penances, from " sram," to work hard, etc. When it became the name of the Buddhist ascetics, the language had changed, and " sramana " was pronounced " samana." Now there is another Sanskrit root. " sam," to quiet, which in Pâli becames likewise " sam," and from this root " sam," to quiet, and not from " sram," to tire, did the popular etymology of the day and the writer of our verse derive the title of the Buddhist priests. The original form " sramana " became known to the Greeks as Σαρμάναι, that of " samana " as Σαμαναῖοι; the former though Megasthenes, the latter through Bardesanes, 80–60 B. C. See Lassen, *Indische Alterthumskunde*, ii. 700. The Chinese " Shamen " and the Tungusian " Shamen " come from the same source, though the latter is sometimes doubted.

266.

A man is not a mendicant (Bhikshu), simply because he asks others for alms; he who adopts the whole law is a Bhikshu, not he who only begs.

267.

He who is above good and evil, who is chaste, who with knowledge passes through the world, he indeed is called a Bhikshu.

268, 269.

A man is not a Muni because he observes silence (mona, *i. e.* mauna), if he is foolish and ignorant; but the wise who, taking the balance, chooses the good and avoids evil, he is a " Muni," and is a " Muni " thereby; he who in this world weighs both sides is called a " Muni."

270.

A man is not an Elect (Ariya) because he injures living creatures; because he has pity on all living creatures, therefore is a man called " Ariya."

271, 272.

Not only by discipline and vows, not only by much learning, not by entering into a trance, not by sleep-

---

(266, 270.) The etymologies here given of the ordinary titles of the followers of Buddha are entirely fanciful, and are curious only as showing how the people who spoke Pâli had lost the etymological consciousness of their language. A "Bhikshu" is a beggar, *i. e.*, a Buddhist friar who has left his family and lives entirely on alms. "Muni" is a sage, hence "Sâkya-muni," the name of Gautama. "Muni" comes from "man," to think, and from "muni" comes "mauna," silence. "Ariya," again, is the general name of those who embrace a religious life. It meant originally "respectable, noble." In v. 270 it seems as if the writer wished to guard against deriving "ariya" from "ari," enemy. See note to v. 22.

ing alone, do I learn the happiness of release which no worldling can know. A Bhikshu receives confidence when he has reached the complete destruction of all desires!

(272.) The last line is obscure, because the commentary is imperfect.

CHAPTER XX.

THE WAY.

273.

THE best of ways is the Eightfold; the best of truths the Four Words; the best of virtues passionlessness; the best of men he who has eyes to see.

274.

This is the way, there is no other that leads to the purifying of intelligence. Go ye on this way! Everything else is the deceit of Mâra (the tempter).

275.

If you go on this way, you will make an end of pain! The way was preached by me, when I had understood the removal of the thorns (in the flesh).

(273.) The eightfold or eight-membered way is the technical term for the way by which Nirvâna is attained. See Burnouf, *Lotus*, 519. This very way constitutes the fourth of the Four Truths, or the four words of truth, namely, Du*h*kha, pain; Samudaya, origin; Nirodha, destruction; Mârga, road. *Lotus*, p. 517. See note to v. 178. For another explanation of the Mârga, or way, see Hardy, *Eastern Monachism*, p. 280.

(275.) The "salyas," arrows or thorns, are the "sokasalya," the arrows of grief. Buddha himself is called "mahasalya-hartâ," the great remover of thorns. *Lalita-vistara*, p. 550; *Mahâbh.* xii. 5616.

276.

You yourself must make an effort. The Tathâgatas (Buddhas) are only preachers. The thoughtful who enter the way are freed from the bondage of Mâra.

277.

"All created things perish," he who knows and sees this becomes passive in pain; this is the way to purity.

278.

"All creatures are grief and pain," he who knows and sees this becomes passive in pain; this is the way to purity.

279.

"All forms are unreal," he who knows and sees this becomes passive in pain; this is the way to purity.

280.

He who does not rise when it is time to rise, who, though young and strong, is full of sloth, whose will and thought are weak, that lazy and idle man will never find the way to knowledge.

281.

Watching his speech, well restrained in mind, let a man never commit any wrong with his body! Let a man but keep these three roads of action clear, and he will achieve the way which is taught by the wise.

282.

Through zeal knowledge is gotten, through lack of

(277.) See v. 255.
(278.) See v. 203.
(279.) "Dhamma" is here explained, like "samkhâra," as the five "khandha," *i. e.*, as what constitutes a living body.

zeal knowledge is lost; let a man who knows this double path of gain and loss thus place himself that knowledge may grow.

283.

Cut down the whole forest of lust, not the tree! From lust springs fear. When you have cut down every tree and every shrub, then, Bhikshus, you will be free!

284.

So long as the love of man towards women, even the smallest, is not destroyed, so long is his mind in bondage, as the calf that drinks milk is to its mother.

285.

Cut out the love of self, like an autumn lotus, with thy hand! Cherish the road of peace. Nirvâna has been shown by Sugata (Buddha).

286.

"Here I shall dwell in the rain, here in winter and summer," thus meditates the fool, and does not think of his death.

287.

Death comes and carries off that man, surrounded by children and flocks, his mind distracted, as a flood carries off a sleeping village.

(282.) "Bhûri" was rightly translated "intelligentia" by Dr. Fausböll. Dr. Weber renders it by "Gedeihen," but the commentator distinctly explains it as "vast knowledge," and in the technical sense the word occurs after "vidyà" and before "midhà," in the *Lalita-vistara*, p. 541.

(283.) A pun, "vana" meaning both "lust" and "forest."

(286.) "Antarâya," according to the commentator, "gîvitàntarâya," *i. e.*, interitus, death. In Sanskrit, "antarita" is used in the sense of "vanished" or "perished."

(287.) See notes to v. 47, and cf. *Mahâbh*. xii. 9944, 6540.

288.

Sons are no help, nor a father, nor relations; there is no help from kinsfolk for one whom Death has seized.

289.

A wise and good man who knows the meaning of this, should quickly clear the way that leads to Nirvâna.

CHAPTER XXI.

MISCELLANEOUS.

290.

IF by leaving a small pleasure one sees a great pleasure, let a wise man leave the small pleasure, and look to the great.

291.

He who, by causing pain to others, wishes to obtain pleasure himself, he, entangled in the bonds of hatred, will never be free from hatred.

292.

What ought to be done is neglected, what ought not to be done is done; the sins of unruly, thoughtless people are always increasing.

293.

But they whose whole watchfulness is always directed to their body, who do not follow what ought not to be done, and who steadfastly do what ought to be done, the sins of such watchful and wise people will come to an end.

294.

A true Brâhma*n*a, though he has killed father and mother, and two valiant kings, though he has destroyed a kingdom with all its subjects, is free from guilt.

295.

A true Brâhma*n*a, though he has killed father and mother, and two holy kings, and even a fifth man, is free from guilt.

296.

The disciples of Gotama (Buddha) are always well awake, and their thoughts day and night are always set on Buddha.

297.

The disciples of Gotama are always well awake, and their thoughts day and night are always set on the Law.

298.

The disciples of Gotama are always well awake, and their thoughts day and night are always set on the Church.

299.

The disciples of Gotama are always well awake, and their thoughts day and night are always set on their body.

300.

The disciples of Gotama are always well awake, and their mind day and night always delights in compassion.

301.

The disciples of Gotama are always well awake, and their mind day and night always delights in meditation.

302.

The hard parting, the hard living alone, the unin-

(294, 295.) These two verses are either meant to show that a truly holy man who by accident commits all these crimes is guiltless, or they refer to some particular event in Buddha's history. The commentator is so startled that he explains them allegorically. The meaning of "veyyaggha" I do not understand.

habitable houses are painful; painful is the company with men who are not our equals; subject to pain are the travelling friars; therefore let no man be a travelling friar, and he will not be subject to pain.

303.

Whatever place a faithful, virtuous, celebrated, and wealthy man chooses, there he is respected.

304.

Good people shine from afar, like the snowy mountains; bad people are not seen, like arrows shot by night.

(302.) Unless this verse formed part of a miscellaneous chapter, I should hardly have ventured to translate it as I have. If the verse means anything, it means that parting with one's friends, living in the wilderness, or in wretched hovels, or travelling about from place to place, homeless and dependent on casual charity, is nothing but pain and grief, and, we should say, according to the author's opinion, useless. In other verses, on the contrary, this very life, this parting with all one holds dear, living in solitude, and depending on alms, is represented as the only course that can lead a man to wisdom, peace, and Nirvâna. Such contradictions, strange as they sound, are not uncommon in the literature of the Brahmans. Here, too, works are frequently represented as indispensable to salvation, and yet, in other places, and from a higher point of view, these very works are condemned as useless, nay, even as a hindrance in a man's progress to real perfection. It is possible that the same view found advocates even in the early days of Buddhism, and that, though performing the ordinary duties, and enjoying the ordinary pleasures of life, a man might consider that he was a truer disciple of Buddha than the dreamy inhabitant of a Vihâra, or the mendicant friar who every morning called for alms at the layman's door (cf. vv. 141, 142). The next verse confirms the view which I have taken.

Should it not be "asamânasamvâso," *i. e.*, living with people who are not one's equals, which was the case in the Buddhist communities, and must have been much against the grain of the Hindus, accustomed, as they were, to live always among themselves, among their own relations, their own profession, their own caste? Living with his superiors is equally disagreeable to a Hindu, as living with his inferiors "Asamâma," unequal, might easily be mistaken for "samâna," proud

305.

He who, without ceasing, practices the duty of eating alone and sleeping alone, he, subduing himself, alone will rejoice in the destruction of all desires, as if living in a forest.

(305.) I have translated this verse so as to bring it into something like harmony with the preceding verses. "Vanânte," according to a pun pointed out before (v. 283), means both "in the end of a forest," and "in the end of desires."

CHAPTER XXII.

THE DOWNWARD COURSE.

306.

HE who says what is not, goes to hell; he also who, having done a thing, says I have not done it. After death both are equal, they are men with evil deeds in the next world.

307.

Many men whose shoulders are covered with the orange gown are ill-conditioned and unrestrained; such evil-doers by their evil deeds go to hell.

308.

Better it would be to swallow a heated iron ball, like flaring fire, than that a bad unrestrained fellow should live on the charity of the land.

309.

Four things does a reckless man gain who covets

(306.) I translate "niraya," the exit, the downward course, the evil path, by "hell," because the meaning assigned to that ancient mythological name by Christian writers comes so near to the Buddhist idea of "niraya," that it is difficult not to believe in some actual contact between these two streams of thought. See also *Mahábh.* xii. 7176. "Abhûtavâdin" is mentioned as a name of Buddha, "sarvasamskârapratisuddhatvât." *Lalita-vistara*, p. 555.

(308.) The charity of the land, *i. e.*, the alms given, from a sense of religious duty, to every mendicant that asks for it.

his neighbor's wife, — a bad reputation, an uncomfortable bed, thirdly, punishment, and lastly, hell.

310.

There is bad reputation, and the evil way (to hell); there is the short pleasure of the frightened in the arms of the frightened, and the king imposes heavy punishment; therefore let no man think of his neighbor's wife.

311.

As a grass-blade, if badly grasped, cuts the arm, badly-practiced asceticism leads to hell.

312.

An act carelessly performed, a broken vow, and hesitating obedience to discipline, all this brings no great reward.

313.

If anything is to be done, let a man do it, let him attack it vigorously! A careless pilgrim only scatters the dust of his passions more widely.

314.

An evil deed is better left undone, for a man repents of it afterwards; a good deed is better done, for having done it, one does not repent.

(309, 310.) The four things mentioned in verse 309 seem to be repeated in verse 310. Therefore, "apuññalâbha," bad fame, is the same in both: "gatî pâpikâ" must be "niraya;" "da*n*da" must be "nindâ," and "ratî thokikâ" explains the "anikâmaseyya*m*." Buddhaghosha takes the same view of the meaning of "anikâmaseyya," i. e., "yathâ i*kkh*ati eva*m* seyyam alabhitvâ, ani*kkh*ita*m* parittakam eva kâla*m* seyya*m* labhati," not obtaining the rest as he wishes it, he obtains it, as he does not wish it, i. e., for a short time only.

(313.) As to "raga" meaning "dust" and "passion," see *Parables*, pp. 65 and 66.

315.

Like a well-guarded frontier fort, with defenses within and without, so let a man guard himself. Not a moment should escape, for they who allow the right moment to pass, suffer pain when they are in hell.

316.

They who are ashamed of what they ought not to be ashamed of, and are not ashamed of what they ought to be ashamed of, such men, embracing false doctrines, enter the evil path.

317.

They who fear when they ought not to fear, and fear not when they ought to fear, such men, embracing false doctrines, enter the evil path.

318.

They who forbid when there is nothing to be forbidden, and forbid not when there is something to be forbidden, such men, embracing false doctrines, enter the evil path.

319.

They who know what is forbidden as forbidden, and what is not forbidden as not forbidden, such men, embracing the true doctrine, enter the good path.

CHAPTER XXIII.

THE ELEPHANT.

320.

SILENTLY shall I endure abuse as the elephant in battle endures the arrow sent from the bow: for the world is ill-natured.

321.

A tamed elephant they lead to battle, the king mounts a tamed elephant; the tamed is the best among men, he who silently endures abuse.

322.

Mules are good, if tamed, and noble Sindhu horses, and elephants with large tusks; but he who tames himself is better still.

323.

For with these animals does no man reach the untrodden country (Nirvâna), where a tamed man goes on a tamed animal, namely on his own well-tamed self.

(320.) The elephant is with the Buddhists the emblem of endurance and self-restraint. Thus Buddha himself is called "Nâga," the Elephant (*Lalita-vistara*, p. 553), or "Mahânâga," the great Elephant (*Lalita-vistara*, p. 553), and in one passage (*Lalita-vistara*, p. 554) the reason of this name is given, by stating that Buddha was "sudânta," well-tamed, like an elephant.

Cf. Manu, vi. 47, "ativâdâ*m*s titiksheta."

(323.) I read, as suggested by Dr. Fausböll, "yath' attanâ sudantena danto dantena ga*kkh*ati." Cf. v. 160. The India Office MS. reads "na hi etchi *th*ânchi ga*kkh*eya agata*m* disam, yath' attâna*m* sudantena danto dantena ga*kkh*ati." As to "*th*ânchi" instead of yânchi," see v. 224.

324.

The elephant called Dhamapâlaka, his temples running with sap, and difficult to hold, does not eat a morsel when bound; the elephant longs for the elephant grove.

325.

If a man becomes fat and a great eater, if he is sleepy and rolls himself about, that fool, like a hog fed on wash, is born again and again.

326.

This mind of mine went formerly wandering about as it liked, as it listed, as it pleased; but I shall now hold it in thoroughly, as the rider who holds the hook holds in the furious elephant.

327.

Be not thoughtless, watch your thoughts! Draw yourself out of the evil way, like an elephant sunk in mud.

328.

If a man find a prudent companion who walks with him, is wise, and lives soberly, he may walk with him, overcoming all dangers, happy, but considerate.

(326.) "Yoniso," *i. e.*, "yonisah," is rendered by Dr. Fausböll "sapientiâ," but the reference which he gives to Hemakandra (ed. Bochtlingk and Rieu, p. 281) shows clearly that it meant "origin," or "cause." "Yoniso" occurs frequently as a mere adverb, meaning thoroughly, radically (*Dhammap.* p. 359), and "yoniso manasikâra" (*Dhammap.* p. 110) means "taking to heart" or "minding thoroughly." In the *Lalita-vistara*, p. 41, the commentator has clearly mistaken "yonisah," changing it to "yesniso," and explaining it by "yamanisam," whereas M. Foucaux has rightly translated it by "depuis l'origine." Professor Weber imagines he has discovered in "yonisâh" a *double-entendre*, but even grammar would show that our author is innocent of it.

329.

If a man find no prudent companion who walks with him, is wise, and lives soberly, let him walk alone, like a king who has left his conquered country behind, — like a lonely elephant.

330.

It is better to live alone, there is no companionship with a fool; let a man walk alone, let him commit no sin, with few wishes, like the lonely elephant.

331.

If an occasion arises, friends are pleasant; enjoyment is pleasant if it is mutual; a good work is pleasant in the hour of death; the giving up of all grief is pleasant.

332.

Pleasant is the state of a mother, pleasant the state of a father, pleasant the state of a Sramana, pleasant the state of a Brâhmana.

333.

Pleasant is virtue lasting to old age, pleasant is a faith firmly rooted; pleasant is attainment of intelligence, pleasant is avoiding of sins.

(332.) The commentator throughout takes these words, like "matteyyatâ," etc., to signify, not the status of a mother, or maternity, but reverence shown to a mother.

CHAPTER XXIV.

THIRST.

334.

THE thirst of a thoughtless man grows like a creeper; he runs hither and thither, like a monkey seeking fruit in the forest.

335.

Whom this fierce thirst overcomes, full of poison, in this world, his sufferings increase like the abounding Bîra*n*a grass.

336.

He who overcomes this fierce thirst, difficult to be conquered in this world, sufferings fall off from him, like water-drops from a lotus leaf.

337.

This salutary word I tell you, as many as are here come together: " Dig up the root of thirst, as he who wants the sweet-scented Usîra root must dig up the Bîra*n*a grass, that Mâra (the tempter) may not crush you again and again, as the stream crushes the reeds.'

338.

As a tree is firm as long as its root is safe, and grows again even though it has been cut down, thus,

---

(335.) Vîrana grass is the *Andropogon muricatum*, and the scented root of it is called " usîra " (cf. v. 337).

unless the yearnings of thirst are destroyed, this pain (of life) will return again and again.

339.

He whose desire for pleasure runs strong in the thirty-six channels, the waves will carry away that misguided man, namely, his desires which are set on passion.

340.

The channels run everywhere, the creeper (of passion) stands sprouting; if you see the creeper springing up, cut its root by means of knowledge.

341.

A creature's pleasures are extravagant and luxurious; sunk in lust and looking for pleasure, men undergo (again and again) birth and decay.

342.

Men, driven on by thirst, run about like a snared hare; held in fetters and bonds, they undergo pain for a long time, again and again.

343.

Men, driven on by thirst, run about like a snared

(338.) On "Anusaya," *i. e.*, "anusaya," see Wassiljew, *Der Buddhismus*, p. 240 *seq.*

(339.) The thirty-six channels, or passions, which are divided by the commentator into eighteen external and eighteen internal, are explained by Burnouf (*Lotus*, p. 649), from a gloss of the "*G*inaalamkâra: " " L'indication précise des affections dont un Buddha acte indépendant, affections qui sont au nombre de dix-huit, nous est fourni par la glose d'un livre appartenant aux Buddhistes de Ceylan," etc.

"Vâhâ," which Dr. Fausböll translates by "equi," may be "vahâ," undæ.

hare; let therefore the mendicant who desires passionlessness for himself, drive out thirst!

344.

He who in a country without forests (*i. e.*, after having reached Nirvâ*n*a) gives himself over to forest-life (*i. e.*, to lust), and who, when removed from the forest (*i. e.*, from lust), runs to the forest (*i. e.*, to lust), look at that man! though free, he runs into bondage.

345.

Wise people do not call that a strong fetter which is made of iron, wood or hemp; far stronger is the care for precious stones and rings, for sons and a wife.

346.

That fetter do wise people call strong which drags down, yields, but is difficult to undo; after having cut this at last, people enter upon their pilgrimage, free from cares, and leaving desires and pleasures behind.

347.

Those who are slaves to passions, run up and down the stream (of desires) as a spider runs up and down the web which he has made himself; when they have cut this, people enter upon their pilgrimage, free from cares, leaving desires and pleasures behind.

(344) This verse seems again full of puns, all connected with the twofold meaning of "vana," forest and lust. By replacing "forest" by "lust," we may translate: "He who, when free from lust, gives himself up to lust, who, when removed from lust runs into lust, look at that man," etc. "Nibbana," though with a short a, may be intended to remind the hearer of Nibbâna.

(345.) "Apekhâ, apekshâ," care; see Manu, vi. 41, 49.

(346.) "Paribba*g*," *i. e.* "parivra*g*;" see Manu, vi. 41.

(347.) The commentator explains the simile of the spider as fol

348.

Give up what is before, give up what is behind, give up what is in the middle, when thou goest to the other shore of existence; if thy mind is altogether free, thou wilt not again enter into birth and decay.

349.

If a man is tossed about by doubts, full of strong passions, and yearning only for what is delightful, his thirst will grow more and more, and he will indeed make his fetters strong.

350.

If a man delights in quieting doubts, and, always reflecting, dwells on what is not delightful, he certainly will remove, nay, he will cut the fetter of Mâra.

351.

He who has obtained rest, who does not tremble, who is without thirst and without blemish, he has broken all the thorns of life: this will be his last body.

352.

He who is without thirst and without affection, who understands the words and their interpretation, who knows the order of letters (those which are before and which are after), he has received his last body, he is called the great sage, the great man.

---

lows: "As a spider, after having made its thread-web, sits in the middle or the centre, and after killing with a violent rush a butterfly or a fly which has fallen in its circle, drinks its juice, returns, and sits again in the same place, in the same manner creatures who are given to passions, depraved by hatred, and maddened by wrath, run along the stream of thirst which they have made themselves, and cannot cross it," etc.

(352.) As to "Nirutti," and its technical meaning among the Bud-

353.

"I have conquered all, I know all, in all conditions of life I am free from taint; I have left all, and through the destruction of thirst I am free; having learnt myself, whom shall I teach?"

354.

The gift of the law exceeds all gifts; the sweetness of the law exceeds all sweetness; the delight in the law exceeds all delights; the extinction of thirst overcomes all pain.

355.

Pleasures destroy the foolish, if they look not for the other shore; the foolish by his thirst for pleasures destroys himself, as if he were his own enemy.

356.

The fields are damaged by weeds; mankind is damaged by passion: therefore a gift bestowed on the passionless brings great reward.

357.

The fields are damaged by weeds, mankind is damaged by hatred: therefore a gift bestowed on those who do not hate brings great reward.

---

dhists, see Burnouf, *Lotus*, p. 841. Fausböll translates "niruttis vocabulorum peritus," which may be right. Could not "sannipâta" mean "saṃhitâ" or "sannikarsha"? "Sannipâta" occurs in the *S*âkala-prîtisâkhya, but with a different meaning.

(354.) The "dhammadâna," or gift of the law, is the technical term for instruction in the Buddhist religion. See *Parables*, p. 160, where the story of the "Sakkadevarâga" is told, and where a free rendering of our verse is given.

358.

The fields are damaged by weeds, mankind is damaged by vanity: therefore a gift bestowed on those who are free from vanity brings great reward.

359.

The fields are damaged by weeds, mankind is damaged by wishing: therefore a gift bestowed on those who are free from wishes brings great reward.

CHAPTER XXV.

THE BHIKSHU (MENDICANT).

360.

RESTRAINT in the eye is good. good is restraint in the ear, in the nose restraint is good, good is restraint in the tongue.

361.

In the body restraint is good, good is restraint in speech, in thought restraint is good, good is restraint in all things. A Bhikshu, restrained in all things, is freed from all pain.

362.

He who controls his hand, he who controls his feet, he who controls his speech, he who is well controlled, he who delights inwardly, who is collected, who is solitary and content, him they call Bhikshu.

(362.) "Agghattarata," *i. e.,* "adhyâtmarata," is an expression which we may take in its natural sense, in which case it would simply mean, delighting inwardly. But "adhyâtmarata" has a technical sense in Sanskrit and with the Brahmans. They use it in the sense of delighting in the Adhyâtman, *i. e.,* the Supreme Self, or Brahman See Manu, vi. 49, and Kullûka's commentary. As the Buddhists do not recognize a Supreme Self or Brahman, they cannot use the word in its Brahmanical sense, and thus we find that Buddhaghosha explains it as "delighting in meditation on the Kammasthâna, a Buddhist formulary, whether externally or internally." I am not certain of the exact meaning of Buddhaghosha's words, but whatever they mean, it is quite clear that he does not take "adhyâtmarata" in the Brahmanical sense. The question then arises who used the

363.

The Bhikshu who controls his mouth, who speaks wisely and calmly, who teaches the meaning and the Law, his word is sweet.

364.

He who dwells in the Law, delights in the Law, meditates on the Law, follows the Law, that Bhikshu will never fall away from the true Law.

365.

Let him not despise what he has received, nor ever envy others: a mendicant who envies others does not obtain peace of mind.

366.

A Bhikshu who, though he receives little, does not despise what he has received, even the gods will praise him, if his life is pure, and if he is not slothful.

367.

He who never identifies himself with his body and

term first, and who borrowed it, and here it would seem, considering the intelligible growth of the word in the philosophical systems of the Brahmans, that the priority belongs for once to the Brahmans.

(363.) On "artha" and "dharma," see Stanislas Julien, *Les Avadânas*, i. 217, note: "Les quatre connaissances sont; 1º la connaissance du sens (artha) ; 2º la connaissance de la Loi (dharma); 3º la connaissance des explications (niroukti); 4' la connaissance de l'intelligence (prâtibhâna)."

(364.) The expression "dhammârâmo," having his garden or delight (Lustgarten) in the Law, is well matched by the Brahmanic expression "ekârâma," *i. e.*, "nirdvandva." *Mahâbh.* xiii. 1930.

(367.) "Nâmarûpa" is here used again in its technical sense of body and soul, neither of which is "âtman," or self. "Asat," what is not, may therefore mean the same as "nâmarûpa," or we may take it in the sense of what is no more, as, for instance, the beauty or youth of the body, the vigor of the mind, etc.

soul, and does not grieve over what is no more, he indeed is called a Bhikshu.

368.

The Bhikshu who acts with kindness, who is calm in the doctrine of Buddha, will reach the quiet place (Nirvâna), cessation of natural desires, and happiness.

369.

O Bhikshu, empty this boat! if emptied, it will go quickly; having cut off passion and hatred, thou wilt go to Nirvâna.

370.

Cut off the five (senses), leave the five, rise above the five? A Bhikshu, who has escaped from the five fetters, he is called Oghatinna, "Saved from the flood."

371.

Meditate, O Bhikshu, and be not heedless! Do not direct thy thought to what gives pleasure! that thou mayest not for thy heedlessness have to swallow the iron ball (in hell), and that thou mayest not cry out when burning, "This is pain."

372.

Without knowledge there is no meditation, without meditation there is no knowledge: he who has knowledge and meditation is near unto Nirvâna.

(371.) The swallowing of hot iron balls is considered as a punishment in hell; see v. 308. Professor Weber has perceived the right meaning of "bhavassu," which can only be "bhâvayasva," but I doubt whether the rest of his rendering is right, "Do not swallow by accident an iron ball."

373.

A Bhikshu who has entered his empty house, and whose mind is tranquil, feels a more than human delight when he sees the law clearly.

374.

As soon as he has considered the origin and destruction of the elements (khandha) of the body, he finds happiness and joy which belong to those who know the immortal (Nirvâna).

375.

And this is the beginning here for a wise Bhikshu: watchfulness over the senses, contentedness, restraint under the Law; keep noble friends whose life is pure, and who are not slothful. ·

376.

Let him live in charity, let him be perfect in his duties; then in the fullness of delight he will make an end of suffering.

377.

As the Vassikâ-plant sheds its withered flowers, men should shed passion and hatred, O ye Bhikshus!

378.

The Bhikshu whose body and tongue and mind are quieted, who is collected, and has rejected the baits of the world, he is called Quiet.

379.

Rouse thyself by thyself, examine thyself by thyself, thus self-protected and attentive wilt thou live happily, O Bhikshu!

380.

For self is the lord of self, self is the refuge of self, therefore curb thyself as the merchant curbs a good horse.

381.

The Bhikshu, full of delight, who is calm in the doctrine of Buddha will reach the quiet place (Nirvâna), cessation of natural desires and happiness.

382.

He who, even as a young Bhikshu, applies himself to the doctrine of Buddha, brightens up this world, like the moon when free from clouds.

(381.) See verse 368.

CHAPTER XXVI.

THE BRÂHMAṆA.

383.

STOP the stream valiantly, drive away the desires, O Brâhmaṇa! When you have understood the destruction of all that was made, you will understand that which was not made.

384.

If the Brâhmaṇa has reached the other shore in both laws (in restraint and contemplation), all bonds vanish from him who has obtained knowledge.

385.

He for whom there is neither this nor that shore, nor both, him, the fearless and unshackled, I call indeed a Brâhmaṇa.

386.

He who is thoughtful, blameless, settled, dutiful, without passions, and who has attained the highest end, him I call indeed a Brâhmaṇa.

387.

The sun is bright by day, the moon shines by night, the warrior is bright in his armor, the Brâhmaṇa is

(385.) The exact meaning of the two shores is not quite clear and the commentator who takes them in the sense of internal and external organs of sense, can hardly be right. See v. 86.

bright in his meditation; but Buddha, the Awakened, is bright with splendor day and night.

388.

Because a man is rid of evil, therefore he is called Brâhmana; because he walks quietly, therefore he is called *S*ramana; because he has sent away his own impurities, therefore he is called Pravra*g*ita (a pilgrim).

389.

No one should attack a Brâhmana, but no Brâhmana (if attacked) should let himself fly at his aggressor! Woe to him who strikes a Brâhmana, more woe to him who flies at his aggressor!

390.

It advantages a Brâhmana not a little if he holds his mind back from the pleasures of life; when all wish to injure has vanished, pain will cease.

391.

Him I call indeed a Brâhmana who does not offend by body, word, or thought, and is controlled on these three points.

(388.) These would-be etymologies are again interesting as showing the decline of the etymological life of the spoken language of India at the time when such etymologies became possible. In order to derive "Brâhma*n*a" from "vâh," it must have been pronounced "bâhmano;" "vâh," to remove, occurs frequently in the Buddhistical Sanskrit. Cf. *Lalita-vistara*, p. 551, l. 1; 553, l. 7. See note to verse 265.

(390.) I am afraid I have taken too much liberty with this verse. Dr. Fausböll translates: "Non Brâhmanæ hoc paulo melius, quando retentio fit mentis a jucundis." In the second verse he translates "hi*m*samano," or "hi*m*samano," by "violenta mens;" Dr. Weber by "der Geist der Schadsucht." Might it be "hi*m*syamâna*h*," injured, and "nivattati," he is quiet, patient? "Ahimsâmana*h*" would be, with the Buddhists, the spirit of love. Luke xi. 39.

392.

After a man has once understood the Law as taught by the Well-awakened (Buddha), let him worship it carefully, as the Brâhmana worships the sacrificial fire.

393.

A man does not become a Brâhmana by his platted hair, by his family, or by both; in whom there is truth and righteousness, he is blessed, he is a Brâhmana.

394.

What is the use of platted hair, O fool! what of the raiment of goatskins? Within thee there is ravening, but the outside thou makest clean.

395.

The man who wears dirty raiments, who is emaciated and covered with veins, who lives alone in the forest, and meditates, him I call indeed a Brâhmana.

396.

I do not call a man a Brâhmana because of his origin or of his mother. He may be called "Sir," and may be wealthy: but the poor, who is free from all attachments, him I call indeed a Brâhmana.

397.

He who has cut all fetters, and who never trembles,

(394.) I have not copied the language of the Bible more than I was justified in. The words are "abbhantaran te gahanam, bâhiram parimaggasi," interna est abyssus, externum mundas.

(395.) The expression "Kisan dhamanisanthatam," is the Sanskrit "krisam dhamanîsantatam," the frequent occurrence of which in the Mahâbhârata has been pointed out by Boehtlingk, s. v. dhamani. It looks more like a Brâhmanic than like a Buddhist phrase.

he who is independent and unshackled, him I call indeed a Brâhmana.

398.

He who has cut the girdle and the strap, the rope with all that pertains to it, he who has burst the bar, and is awakened, him I call indeed a Brâhmana.

399.

He who, though he has committed no offense, endures reproach, bonds, and stripes, him, strong in endurance and powerful, I call indeed a Brâhmana.

400.

He who is free from anger, dutiful, virtuous, without weakness, and subdued, who has received his last body, him I call indeed a Brâhmana.

401.

He who does not cling to pleasures, like water on a lotus leaf, like a mustard seed on the point of an awl, him I call indeed a Brâhmana.

402.

He who, even here, knows the end of his suffering, has put down his burden, and is unshackled, him I call indeed a Brâhmana.

403.

He whose knowledge is deep, who possesses wisdom, who knows the right way and the wrong, who has

---

(399.) The exact meaning of "balânika" is difficult to find. Does it mean, possessed of a strong army, or facing a force, or leading a force? The commentary alone could help us to decide.

attained the highest end, him I call indeed a Brâhmana.

404.

He who keeps aloof both from laymen and from mendicants, goes to no house to beg, and whose desires are small, him I call indeed a Brâhmana.

405.

He who finds no fault with other beings, whether weak or strong, who does not kill nor cause slaughter, him I call indeed a Brâhmana.

406.

He who is tolerant with the intolerant, mild with fault-finders, free from passion among the passionate, him I call indeed a Brâhmana.

407.

He from whom anger and hatred, pride and envy have dropt like a mustard seed from the point of an awl, him I call indeed a Brâhmana.

408.

He who utters true speech, instructive and free from harshness, so that he offend no one, him I call indeed a Brâhmana.

(404.) "Anokasâri" is translated by Dr. Fausböll "sine domicilio grassantem;" by Dr. Weber, "ohne Heim wandelt." The commentator seems to support my translation. He says that a man who has no intercourse either with householders or with those who have left their houses, but may still dwell together in retirement from the world, is "anilayakara," *i. e.*, a man who goes to nobody's abode, in order to see, to hear, to talk, or to eat. He then explains "anokasârin" by the same word, "anâlayakârin," *i. e.*, a man who goes to nobody's residence for any purpose, — and in our case, I suppose, principally not for the purpose of begging.

409.

He who takes nothing in the world that is not given him, be it long or short, small or large, good or bad, him I call indeed a Brâhmana.

410.

He who fosters no desires for this world or for the next, has no inclinations, and is unshackled, him I call indeed a Brâhmana.

411.

He who has no interests, and when he has understood (the truth), does not say How, how? — he who can dive into the Immortal, him I call indeed a Brâhmana.

412.

He who is above good and evil, above the bondage of both, free from grief, from sin, from impurity, him I call indeed a Brâhmana.

413.

He who is bright like the moon, pure, serene, and undisturbed, in whom all gayety is extinct, him I call indeed a Brâhmana.

(411.) "Akathamkathi" is explained by Buddhaghosha as meaning, free from doubt or hesitation. He also uses "kathamkathâ" in the sense of doubt (verse 414). In the *Kâvyâdarsa*, iii. 17, the commentator explains "akatham" by "kathârahitam, nirvivâdam," which would mean, without a "kathâ," a speech, a story without contradiction, unconditionally. From our passage, however, it seems as if "kathamkathâ" was a noun derived from "kathamthayati," to say How, how? so that neither the first nor the second element had anything to do with "kath," to relate; and in that case "akatham," too, ought to be taken in the sense of "without a Why."

(412.) See verse 39. The distinction between good and evil vanishes when a man has retired from the world, and has ceased to act, longing only for deliverance.

414.

He who has traversed this mazy, impervious world and its vanity, who is through, and has reached the other shore, is thoughtful, guileless, free from doubts, free from attachment, and content, him I call indeed a Brâhmana.

415.

He who, leaving all desires, travels about without a home, in whom all concupiscence is extinct, him I call indeed a Brâhmana.

416.

He who, leaving all longings, travels about without a home, in whom all covetousness is extinct, him I call indeed a Brâhmana.

417.

He who, after leaving all bondage to men, has risen above all bondage to the gods, who is free from every bondage, him I call indeed a Brâhmana.

418.

He who has left what gives pleasure and what gives pain, is cold, and free from all germs (of renewed life), the hero who has conquered all the worlds, him I call indeed a Brâhmana.

419.

He who knows the destruction and the return of creatures everywhere, who is free from bondage, welfaring (Sugata), and awakened (Buddha), him I call indeed a Brâhmana.

420.

He whose way the gods do not know, nor spirits

(Gandharvas), nor men, and whose passions are extinct, him, the venerable, I call indeed a Brâhmana.

421.

He who calls nothing his own, whether it be before, behind, or between, who is poor, and free from the love of the world, him I call indeed a Brâhmana.

422.

The manly, the noble, the hero, the great sage, the conqueror, the guileless, the master, the awakened, him I call indeed a Brâhmana.

423.

He who knows his former abodes, who sees heaven and hell, has reached the end of births, is perfect in knowledge and a sage, he whose perfections are all perfect, him I call indeed a Brâhmana.

www.ingramcontent.com/pod-product-compliance
Lightning Source LLC
Chambersburg PA
CBHW022110230426
43672CB00008B/1328